Analytical Techniques in Telecommunications

Edited by
Fraidoon Mazda
MPhil DFH CEng FIEE

With specialist contributions

Focal Press
An imprint of Butterworth-Heinemann
Linacre House, Jordan Hill, Oxford OX2 8DP
A division of Reed Educational and Professional Publishing Ltd

 A member of the Reed Elsevier plc group

OXFORD BOSTON JOHANNESBURG
NEW DELHI SINGAPORE MELBOURNE

First published 1996

© Reed Educational and Professional Publishing Ltd 1996

All rights reserved. No part of this publication
may be reproduced in any material form (including
photocopying or storing in any medium by electronic
means and whether or not transiently or incidentally
to some other use of this publication) without the
written permission of the copyright holder except in
accordance with the provisions of the Copyright,
Designs and Patents Act 1988 or under the terms of a
licence issued by the Copyright Licensing Agency Ltd,
90 Tottenham Court Road, London, England, W1P 9HE.
Applications for the copyright holder's written permission
to reproduce any part of this publication should be addressed
to the publishers

British Library Cataloguing in Publication Data
Mazda, Fraidoon F
   Analytical techniques in telecommunications
   I. Title
   621.382

D
621.382
TEL

ISBN 02405 1451 3

Library of Congress Cataloguing in Publication
Mazda, Fraidoon F.
   Analytical techniques in telecommunications/Fraidoon Mazda
   p. cm.
   Includes bibliographical references and index.
   ISBN 02405 1451 3
   1. Telecommunications. I. Title
   TK5101.M37 1993
   621.382–dc20

92-27846
CIP

Printed and bound in Great Britain by
Biddles Ltd, Guildford and King's Lynn

Analytical Techniques in Telecommunications

Contents

Preface	vii
List of contributors	ix
1. Trigonometric and general formulae	1
2. Calculus	21
3. Series and transforms	31
4. Matrices and determinants	42
5. Statistical analysis	49
6. Fourier analysis	78
7. Queuing theory	121
8. Information theory	166
9. Teletraffic theory	197
10. Coding	227
11. Signals and noise	260
Index	301

Preface

Telecommunications, in common with other scientific subjects, is largely based on analytical techniques, which usually involve mathematical analysis. These techniques are often the first to be learnt during an engineering college course and are also the first to be forgotten, as the student moves from the classroom to an engineering work-place. It is only when problems arise, requiring the application of these long-forgotten techniques, that the deficiency is noticed.

The present book collects together those analytical techniques which are most frequently required within telecommunications. It is intended to serve as a refresher as well as providing a ready reference.

Chapters one to six cover general mathematical techniques, such as trigonometric formulae and equations; calculus; series and transforms; matrices and determinants; statistical analysis; and Fourier analysis. The remainder of the book focuses on subjects which are more specific to telecommunications, such as queuing theory; information theory; teletraffic theory, a key element in the analysis of telecommunications systems; the equally important subject of coding; and the analysis of signals and noise.

Eight authors have contributed to this book, all specialists in their field, and the success of the book is largely due to their efforts. The book is also based on selected chapters which were first published in the much larger volume of the *Telecommunications Engineers' Reference Book*.

Fraidoon Mazda
Bishop's Stortford
April 1996

List of contributors

J Barron
BA MA (Cantab)
University of Cambridge
(Chapters 1 to 4)

William J Fitzgerald
University of Cambridge
(Chapters 6 and 8)

R J Gibbens
MA Dip Math Stat PhD
University of Cambridge
(Chapter 9)

Terry Goble
Nortel Ltd
(Chapter 11)

M D Macleod
MA PhD MIEE
University of Cambridge
(Chapter 10)

Fraidoon Mazda
MPhil DFH CEng FIEE
Nortel Ltd
(Chapter 5)

John Price
Nortel Ltd
(Chapter 11)

Phil Whiting
University of Strathclyde
(Chapter 7)

1. Trigonometric and general formulae

1.1 Mathematical signs and symbols

Sign, symbol	Quantity
$=$	equal to
$\neq$	not equal to
$\equiv$	identically equal to
Δ	corresponds to
$\approx$	approximately equal to
$\rightarrow$	approaches
∞	infinity
$<$	smaller than
$>$	larger than
$\leq$	smaller than or equal to
$\geq$	larger than or equal to
$\lvert a \rvert$	magnitude of a
a^n	a raised to the power n
$a^{1/2} \quad \sqrt{a}$	square root of a
$\bar{a} \quad <a>$	mean value of a
$p\,!$	factorial p $(1 \times 2 \times 3 \times \ldots \times p)$
$\binom{n}{p}$	binomial coefficient
$\sum$	sum

2 Mathematical signs and symbols

Sign, symbol	Quantity
$\prod$	product
$f(x)$	function f of the variable x
$[f(x)]_a^b$	$f(b) - f(a)$
Δx	delta x = finite increment of x
δx	delta x = variation of x
$\dfrac{df}{dx}$ $f'(x)$	differential coefficient of $f(x)$ with respect to x
$\dfrac{d^n f}{dx^n}$ $f^{(n)}(x)$	differential coefficient of order n of $f(x)$
$\dfrac{\partial f(x,y,...)}{\partial x}$ $\left(\dfrac{\partial f}{\partial x}\right)_{y...}$	partial differential coefficient of $f(x,y,...)$ with respect of x
df	the total differential of f
$\int f(x)\,dx$	indefinite integral of $f(x)$ with respect to x
$\int_a^b f(x)\,dx$	definite integral of $f(x)$ from $x = a$ to $x = b$
e	base of natural logarithms
e^x $\exp x$	e raised to the power x
$\log_a x$	logarithm to the base a of x
$lg\ x$ $\log x$ $\log_{10} x$	common (Briggsian) logarithm of x
$lb\ x$ $\log_2 x$	binary logarithm of x
$\sin x$	sine of x
$\cos x$	cosine of x
$\tan x$	tangent of x
$\cot x$	cotangent of x
$\sec x$	secant of x

Sign, symbol	Quantity
cosec x	cosecant of x
arcsin x	arc sine of x
arccos x	arc cosine of x
arctan x	arc tangent of x
arccot x	arc cotangent of x
arcsec x	arc secant of x
arccosec x	arc cosecant of x
sinh x	hyperbolic sine of x
cosh x	hyperbolic cosine of x
tanh x	hyberbolic tangent of x
coth x	hyperbolic cotangent of x
sech x	hyperbolic secant of x
cosech x	hyperbolic cosecant of x
arsinh x	inverse hyperbolic sine of x
arcosh x	inverse hyperbolic cosine of x
artanh x	inverse hyperbolic tangent of x
arcoth x	inverse hyperbolic cotangent of x
arsech x	inverse hyperbolic secant of x
arcosech x	inverse hyperbolic cosecant of x
i, j	imaginary unity, $i^2 = -1$
Re z	real part of z
Im z	imaginary part of z
arg z	argument of z
z^*	conjugate of z
$\overline{A}$, A', A^t	transpose of matrix A
A^*	complex conjugate matrix of matrix A

4 Trigonometric formulae

Sign, symbol	Quantity
A^+	Hermitian conjugate matrix of matrix A
A a	vector
$|A|$	magnitude of vector
$A \cdot B$	scalar product
$A \times B$	vector product
∇	differential vector operator
$\nabla \varphi$ grad φ	gradient of φ
$\nabla \cdot A$ div A	divergence of A
$\nabla \times A$ curl A	curl of A
$\nabla^2 \varphi$ $\Delta \varphi$	Laplacian of φ

1.2 Trigonometric formulae

$$\sin^2 A + \cos^2 A = \sin A \, \mathrm{cosec}\, A = 1$$

$$\sin A = \frac{\cos A}{\cot A} = \frac{1}{\mathrm{cosec}\, A} = (1 - \cos^2 A)^{1/2}$$

$$\cos A = \frac{\sin A}{\tan A} = \frac{1}{\sec A} = (1 - \sin^2 A)^{1/2}$$

$$\tan A = \frac{\sin A}{\cos A} = \frac{1}{\cot A}$$

$$1 + \tan^2 A = \sec^2 A$$

$$1 + \cot^2 A = \mathrm{cosec}^2 A$$

$$1 - \sin A = \mathrm{coversin}\, A$$

$$1 - \cos A = \mathrm{versin}\, A$$

$$\tan \tfrac{1}{2}\theta = t; \quad \sin \theta = \frac{2t}{1+t^2}; \quad \cos \theta = \frac{1-t^2}{1+t^2}$$

$$\cot A = \frac{1}{\tan A}$$

$$\sec A = \frac{1}{\cos A}$$

$$\operatorname{cosec} A = \frac{1}{\sin A}$$

$$\cos(A \pm B) = \cos A \cos B \mp \sin A \sin B$$

$$\sin(A \pm B) = \sin A \cos B \pm \cos A \sin B$$

$$\tan(A \pm B) = \frac{\tan A \pm \tan B}{1 \mp \tan A \tan B}$$

$$\cot(A \pm B) = \frac{\cot A \cot B \mp 1}{\cot B \pm \cot A}$$

$$\sin A \pm \sin B = 2 \sin \tfrac{1}{2}(A \pm B) \cos \tfrac{1}{2}(A \mp B)$$

$$\cos A + \cos B = 2 \cos \tfrac{1}{2}(A + B) \cos \tfrac{1}{2}(A - B)$$

$$\cos A - \cos B = 2 \sin \tfrac{1}{2}(A + B) \sin \tfrac{1}{2}(B - A)$$

$$\tan A \pm \tan B = \frac{\sin(A \pm B)}{\cos A \cos B}$$

$$\cot A \pm \cot B = \frac{\sin(B \pm A)}{\sin A \sin B}$$

$$\sin 2A = 2 \sin A \cos A$$

$$\cos 2A = \cos^2 A - \sin^2 A = 2\cos^2 A - 1 = 1 - 2\sin^2 A$$

$$\cos^2 A - \sin^2 B = \cos(A+B)\cos(A-B)$$

$$\tan 2A = \frac{2 \tan A}{1 - \tan^2 A}$$

$$\sin \tfrac{1}{2} A = \left(\frac{1 - \cos A}{2}\right)^{1/2}$$

$$\cos \tfrac{1}{2} A = \pm\left(\frac{1 + \cos A}{2}\right)^{1/2}$$

$$\tan \tfrac{1}{2} A = \frac{\sin A}{1 + \cos A}$$

$$\sin^2 A = \tfrac{1}{2}(1 - \cos 2A)$$

$$\cos^2 A = \tfrac{1}{2}(1 + \cos 2A)$$

$$\tan^2 A = \frac{1 - \cos 2A}{1 + \cos 2A}$$

$$\tan \tfrac{1}{2}(A \pm B) = \frac{\sin A \pm \sin B}{\cos A + \cos B}$$

$$\cot \tfrac{1}{2}(A \pm B) = \frac{\sin A \pm \sin B}{\cos B - \cos A}$$

1.3 Trigonometric values

Angle	$0°$	$30°$	$45°$	$60°$	$90°$	$180°$	$270°$	$360°$
Radians	0	$\pi/6$	$\pi/4$	$\pi/3$	$\pi/2$	π	$3\pi/2$	2π
Sine	0	$\tfrac{1}{2}$	$\tfrac{1}{2}\sqrt{2}$	$\tfrac{1}{2}\sqrt{3}$	1	0	-1	0
Cosine	1	$\tfrac{1}{2}\sqrt{3}$	$\tfrac{1}{2}\sqrt{2}$	$\tfrac{1}{2}$	0	-1	0	1
Tangent	0	$1/\sqrt{3}$	1	$\sqrt{3}$	∞	0	∞	0

1.4 Approximations for small angles

$$\sin \theta = \theta - \frac{\theta^3}{6}$$

$$\cos \theta = 1 - \frac{\theta^2}{2} \quad (\theta \text{ in radians})$$

$$\tan \theta = \theta + \frac{\theta^3}{3}$$

1.5 Solution of triangles

$$\frac{\sin A}{a} = \frac{\sin B}{b} = \frac{\sin C}{c}$$

$$\cos A = \frac{b^2 + c^2 - a^2}{2bc}$$

$$\cos B = \frac{c^2 + a^2 - b^2}{2ca}$$

$$\cos C = \frac{a^2 + b^2 - c^2}{2ab}$$

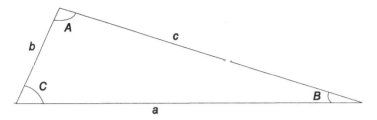

Figure 1.1 Triangle

where A, B, C and a, b, c are shown in Figure 1.1.
If $s = \frac{1}{2}(a + b + c)$:

$$\sin \frac{A}{2} = \sqrt{\frac{(s-b)(s-c)}{bc}}$$
$$\sin \frac{B}{2} = \sqrt{\frac{(s-c)(s-a)}{ca}}$$
$$\sin \frac{C}{2} = \sqrt{\frac{(s-a)(s-b)}{ab}}$$
$$\cos \frac{A}{2} = \sqrt{\frac{s(s-a)}{bc}}$$
$$\cos \frac{B}{2} = \sqrt{\frac{s(s-b)}{ca}}$$
$$\cos \frac{C}{2} = \sqrt{\frac{s(s-c)}{ab}}$$
$$\tan \frac{A}{2} = \sqrt{\frac{(s-b)(s-c)}{s(s-a)}}$$
$$\tan \frac{B}{2} = \sqrt{\frac{(s-c)(s-a)}{s(s-b)}}$$
$$\tan \frac{C}{2} = \sqrt{\frac{(s-a)(s-b)}{s(s-c)}}$$

1.6 Spherical triangle

$$\frac{\sin A}{\sin a} = \frac{\sin B}{\sin b} = \frac{\sin C}{\sin c}$$
$$\cos a = \cos b \cos c + \sin b \sin c \cos A$$

8 Exponential form

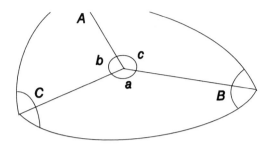

Figure 1.2 Spherical triangle

$\cos b = \cos c \cos a + \sin c \sin a \cos B$
$\cos c = \cos a \cos b + \sin a \sin b \cos C$

where A, B, C and a, b, c are now as in Figure 1.2.

1.7 Exponential form

$$\sin \theta = \frac{e^{i\theta} - e^{-i\theta}}{2i}$$
$$\cos \theta = \frac{e^{i\theta} + e^{-i\theta}}{2}$$
$$e^{i\theta} = \cos \theta + i \sin \theta$$
$$e^{-i\theta} = \cos \theta - i \sin \theta$$

1.8 De Moivre's theorem

$(\cos A + i \sin A)(\cos B + i \sin B) = \cos (A + B) + i \sin (A + B)$

1.9 Euler's relation

$(\cos \theta + i \sin \theta)^n = \cos n\theta + i \sin n\theta = e^{in\theta}$

1.10 Hyperbolic functions

$$\sinh x = \frac{e^x - e^{-x}}{2}$$

$$\cosh x = \frac{e^x + e^{-x}}{2}$$

$$\tan x = \frac{\sinh x}{\cosh x}$$

Relations between hyperbolic functions can be obtained from the corresponding relations between trigonometric functions by reversing the sign of any term containing the product or implied product of two sines, e.g.:

$$\cosh^2 A - \sinh^2 A = 1$$
$$\cosh 2A = 2\cosh^2 A - 1 = 1 + 2\sinh^2 A = \cosh^2 A + \sinh^2 A$$
$$\cosh(A \pm B) = \cosh A \cosh B \pm \sinh A \sinh B$$
$$\sinh(A \pm B) = \sinh A \cosh B \pm \cosh A \sinh B$$
$$e^x = \cosh x + \sinh x$$
$$e^{-x} = \cosh x - \sinh x$$

1.11 Complex variables

If $z = x + iy$, where x and y are real variables, z is a complex variable and is a function of x and y. z may be represented graphically in an Argand diagram (Figure 1.3).

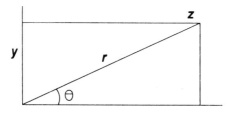

Figure 1.3 Argand diagram

10 Cauchy–Riemann equations

Polar form:

$$z = x + iy = |z|\, e^{i\theta} = |z|(\cos\theta + i\sin\theta)$$
$$x = r\cos\theta \qquad y = r\sin\theta$$
where $r = |z|$

Complex arithmetic:

$$z_1 = x_1 + i\, y_1$$
$$z_2 = x_2 + i\, y_2$$
$$z_1 \pm z_2 = (x_1 \pm x_2) + i\,(y_1 \pm y_2)$$
$$z_1 z_2 = (x_1 x_2 - y_1 y_2) + i\,(x_1 y_2 + x_2 y_1)$$

Conjugate:

$$z^* = x - iy$$
$$z z^* = x^2 + y^2 = |z|^2$$

Function: another complex variable $w = u + iv$ may be related functionally to z by:

$$w = u + iv = f(x + iy) = f(z)$$

which implies:

$$u = u(x,y) \qquad v = v(x,y) \qquad \text{e.g.}$$

$$\cosh z = \cosh(x + iy) = \cosh x \cosh iy + \sinh x \sinh iy$$
$$= \cosh x \cos y + i \sinh x \sin y$$
$$u = \cosh x \cos y \qquad v = \sinh x \sin y$$

1.12 Cauchy–Riemann equations

If $u(x,y)$ and $v(x,y)$ are continuously differentiable with respect to x and y:

$$\frac{\partial u}{\partial x} = \frac{\partial v}{\partial y} \qquad \frac{\partial u}{\partial y} = -\frac{\partial v}{\partial x}$$

$w = f(z)$ is continuously differentiable with respect to z and its derivative is:

$$f'(z) = \frac{\partial u}{\partial x} + i\frac{\partial v}{\partial x} = \frac{\partial v}{\partial y} - i\frac{\partial u}{\partial y} = \frac{1}{i}\left(\frac{\partial u}{\partial y} + i\frac{\partial v}{\partial y}\right)$$

It is also easy to show that $\nabla^2 u = \nabla^2 v = 0$. Since the transformation from z to w is conformal, the curves $u =$ constant and $v =$ constant intersect each other at right angles, so that one set may be used as equipotentials and the other as field lines in a vector field.

1.13 Cauchy's theorem

If $f(z)$ is analytic everywhere inside a region bounded by C and a is a point within C:

$$f(a) = \frac{1}{2\pi i} \int_C \frac{f(z)}{z-a}\, dz$$

This formula gives the value of a fraction at a point in the interior of a closed curve in terms of the values on that curve.

1.14 Zeros, poles and residues

If $f(z)$ vanishes at the point z_0 the Taylor series for z in the region of z_0 has its first two terms zero, and perhaps others also: $f(z)$ may then be written:

$$f(z) = (z - z_0)^n\, g(z)$$

where $g(z_0) \neq 0$. Then $f(z)$ has a zero of order n at z_0. The reciprocal:

12 Some standard forms

$$q(z) = \frac{1}{f(z)} = \frac{h(z)}{(z-z_0)^n}$$

where $h(z) = \dfrac{1}{q(z)} \neq 0$ at z_0. $q(z)$ may be expanded in the form:

$$q(z) = c_{-n}(z-z_0)^n + \ldots + c_{-1}(z-z_0)^{-1} + c_0 + \ldots$$

where c_{-1} is the residue of $q(z)$ at $z = z_0$. From Cauchy's theorem it may be shown that if a function $f(z)$ is analytic throughout a region enclosed by a curve C except at a finite number of poles, the integral of the function around C has a value of $2\pi i$ times the sum of the residues of the function at its poles within C. This fact can be used to evaluate many definite integrals whose indefinite form cannot be found.

1.15 Some standard forms

$$\int_0^{2\pi} e^{\cos\theta}(\cos n\theta - \sin\theta)\,d\theta = \frac{2\pi}{n!}$$

$$\int_0^\infty \frac{x^{a-1}}{1+x}\,dx = \pi\,\operatorname{cosec} a\pi$$

$$\int_0^\infty \frac{\sin\theta}{\theta}\,d\theta = \frac{\pi}{2}$$

$$\int_0^\infty x\exp(-h^2x^2)\,dx = \frac{1}{2h^2}$$

$$\int_0^\infty \frac{x^{a-1}}{1-x}\,dx = \pi\cot a\pi$$

$$\int_0^\infty \exp(-h^2x^2)\,dx = \frac{\sqrt{\pi}}{2h}$$

$$\int_0^\infty x^2\exp(-h^2x^2)\,dx = \frac{\sqrt{\pi}}{4h^3}$$

Trigonometric and general formulae 13

1.16 Coordinate systems

The basic system is the rectangular Cartesian system (x,y,z) to which all other systems are referred. Two other commonly used systems are as follows.

1.16.1 Cylindrical coordinates

Coordinated of point P are (x,y,z) or (r,θ,z), as in Figure 1.4, where:

$x = r \cos \theta \qquad y = r \sin \theta \qquad z = z$

In these coordinates the volume element is $r \, dr \, d\theta \, dz$.

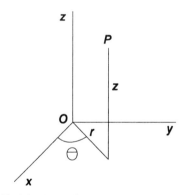

Figure 1.4 Cylindrical coordinates

1.16.2 Spherical polar coordinates

Coordinates of point P are (x,y,z) or (r,θ,φ), as in Figure 1.5, where:

$x = r \sin \theta \cos \varphi \qquad y = r \sin \theta \sin \varphi \qquad z = r \cos \theta$

In these coordinates the volume element is $r^2 \sin \theta \, dr \, d\theta \, d\varphi$.

14 Transformation of integrals

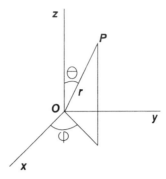

Figure 1.5 Spherical polar coordinates

1.17 Transformation of integrals

$$\iiint f(x,y,z)\,dx\,dy\,dz = \iiint \varphi(u,v,w)\,|J|\,du\,dv\,dw$$

where:

$$J = \begin{vmatrix} \dfrac{\partial x}{\partial u} & \dfrac{\partial y}{\partial u} & \dfrac{\partial z}{\partial u} \\ \dfrac{\partial x}{\partial v} & \dfrac{\partial y}{\partial v} & \dfrac{\partial z}{\partial v} \\ \dfrac{\partial x}{\partial w} & \dfrac{\partial y}{\partial w} & \dfrac{\partial z}{\partial w} \end{vmatrix} = \dfrac{\partial(x,y,z)}{\partial(u,v,w)}$$

is the Jacobian of the transformation of coordinates. For Cartesian to cylindrical coordinates $J = r$, and for Cartesian to spherical polars $J = r^2 \sin\theta$.

1.18 Laplace's equation

The equation satisfied by the scalar potential from which a vector field may be derived by taking the gradient is Laplace's equation, written as:

Trigonometric and general formulae 15

$$\nabla^2 \varphi = \frac{\partial^2 \varphi}{\partial x^2} + \frac{\partial^2 \varphi}{\partial y^2} + \frac{\partial^2 \varphi}{\partial z^2} = 0$$

In cylindrical coordinates:

$$\nabla^2 \varphi = \frac{1}{r} \frac{\delta}{\delta r}\left(r \frac{\delta \varphi}{\delta r}\right) + \frac{1}{r^2} \frac{\delta^2 \varphi}{\delta \theta^2} + \frac{\delta^2 \varphi}{\delta z^2}$$

In spherical polars:

$$\nabla^2 \varphi = \frac{1}{r^2} \frac{\partial}{\partial r}\left(r^2 \frac{\partial \varphi}{\partial r}\right) + \frac{1}{r^2 \sin \theta} \frac{\partial \varphi}{\partial \theta} + \frac{1}{r^2 \sin^2 \theta} \frac{\partial^2 \varphi}{\partial \varphi^2}$$

The equation is solved by setting:

$\varphi = U(u)\, V(u)\, W(w)$

in the appropriate form of the equation, separating the variables and solving separately for the three functions, where (u,v,w) is the coordinate system in use.

In Cartesian coordinates, typically the functions are trigonometric, hyperbolic and exponential; in cylindrical coordinates the function of z is exponential, that of θ trigonometric and that of r is a Bessel function. In spherical polars, typically the function of r is a power of r, that of φ is trigonometric, and that of θ is a Legendre function of $\cos \theta$.

1.19 Solution of equations

1.19.1 Quadratic equation

$$ax^2 + bx + c = 0$$
$$x = -\frac{b}{2a} \pm \frac{\sqrt{b^2 - 4ac}}{2a}$$

16 Solution of equations

In practical calculations if $b^2 > 4ac$, so that the roots are real and unequal, calculate the root of larger modulus first, using the same sign for both terms in the formula, then use the fact that $x_1 x_2 = c/a$ where x_1 and x_2 are the roots. This avoids the severe cancellation of significant digits which may otherwise occur in calculating the smaller root.

For polynomials other than quadratics, and for other functions, several methods of successive approximation are available.

1.19.2 Bisection method

By trial find x_0 and x_1 such that $f(x_0)$ and $f(x_1)$ have opposite sings (see Figure 1.6). Set $x_2 = (x_0 + x_1)/2$ and calculate $f(x_2)$. If $f(x_0) f(x_2)$ is positive, the root lies in the interval (x_1, x_2); if negative in the interval (x_0, x_2); and if zero, x_2 is the root. Continue if necessary using the new interval.

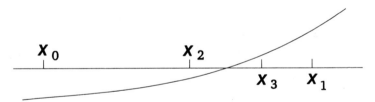

Figure 1.6 Bisection method

1.19.3 Regula falsi

By trial, find x_0 and x_1 as for the bisection method; these two values define two points $(x_0, f(x_0))$ and $(x_1, f(x_1))$. The straight line joining these two points cuts the x axis at the point (see Figure 1.7):

$$x_2 = \frac{x_0 f(x_1) - x_1 f(x_0)}{f(x_1) - f(x_0)}$$

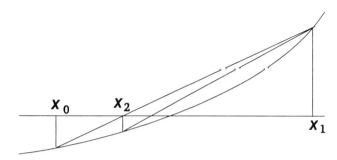

Figure 1.7 Regula falsi

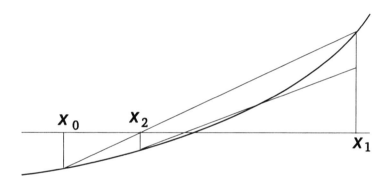

Figure 1.8 Accelerated method

Evaluate $f(x_2)$ and repeat the process for whichever of the intervals (x_0, x_2) or (x_1, x_2) contains the root. This method can be accelerated by halving at each step the function value at the retained end of the interval, as shown in Figure 1.8.

1.19.4 Fixed-point iteration

Arrange the equation in the form:

$x = f(x)$

Choose an initial value of x by trial, and calculate repetitively:

$$x_{k+1} = f(x_k)$$

This process will not always converge.

1.19.5 Newton's method

Calculate repetitively (Figure 1.9):

$$x_{k+1} = x_k - f(x_k)/f'(x_k)$$

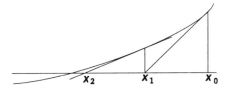

Figure 1.9 Newton's method

This method will converge unless: (a) x_k is near a point of inflexion of the function; or (b) x_k is near a local minimum; or (c) the root is multiple. If one of these cases arises, most of the trouble can be overcome by checking at each stage that:

$$f(x_{k+1}) < f(x_k)$$

and, if not, halving the preceding value of $\left|x_{k+1} - x_k\right|$.

1.20 Method of least squares

To obtain the best fit between a straight line $ax + by = 1$ and several points $(x_1,y_1), (x_2,y_2), ..., (x_n,y_n)$ found by observation, the coefficients a and b are to be chosen so that the sum of the squares of the errors:

$$e_i = ax_i + by_i - 1$$

is a minimum. To do this, first write the set of inconsistent equations:

$$ax_1 + by_1 - 1 = 0$$
$$ax_2 + by_2 - 1 = 0$$
$$\cdot$$
$$\cdot$$
$$\cdot$$
$$ax_n + by_n - 1 = 0$$

Multiply each equation by the value of x it contains and add, obtaining:

$$a \sum_{i=1}^{n} x_i^2 + b \sum_{i=1}^{n} x_i y_i - \sum_{i=1}^{n} x_i = 0$$

Similarly multiply by y and add, obtaining:

$$a \sum_{i=1}^{n} x_i y_i + b \sum_{i=1}^{n} y_i^2 - \sum_{i=1}^{n} y_i = 0$$

Lastly, solve these two equations for a and b, which will be the required values giving the least squares fit.

1.21 Decibels, current and voltage ratio, and power ratio

$$\text{dB} = 10 \log \frac{P_1}{P_2} = 20 \log \frac{V_1}{V_2} = 20 \log \frac{I_1}{I_2}$$

20 Decibels, current and voltage ratio, and power ratio

dB	I_1/I_2 or V_1/V_2	I_2/I_1 or V_2/V_1	P_1/P_2	P_2/P_1
0.5	1.059	0.944	1.122	0.891
1.0	1.122	0.891	1.259	0.794
2.0	1.259	0.794	1.585	0.631
3.0	1.413	0.708	1.995	0.501
4.0	1.585	0.631	2.51	0.398
5.0	1.778	0.562	3.16	0.316
6.0	1.995	0.501	3.98	0.251
7.0	2.24	0.447	5.01	0.200
8.0	2.51	0.398	6.31	0.158
9.0	2.82	0.355	7.94	0.126
10.0	3.16	0.316	10.0	0.100
15.0	5.62	0.178	31.6	0.0316
20.0	10.00	0.100	100	0.0100
25.0	17.78	0.0562	316	0.00316
30.0	31.6	0.0316	1000	0.00100

2. Calculus

2.1 Derivative

$$f'(x) = \lim_{\partial x \to 0} \frac{f(x + \partial x) - f(x)}{\partial x}$$

If u and v are functions of x:

$$\left(\frac{u}{v}\right)' = \frac{u'v - uv'}{v^2}$$

$$(uv)^{(n)} = u^{(n)}v + v^{(n-1)}v^{(1)} + \ldots + {}^nC_p\, u^{(n-p)}v^{(p)} + \ldots + uv^{(n)}$$

where:

$${}^nC_p = \frac{n!}{p!(n-p)!}$$

If $z = f(x)$ and $y = g(z)$, then:

$$\frac{dy}{dx} = \frac{dy}{dz}\frac{dz}{dx}$$

2.2 Maxima and minima

$f(x)$ has a stationary point wherever $f'(x) = 0$: the point is a maximum, minimum or point of inflexion according as $f''(x)$ is less than, greater than or equal to zero. $f(x,y)$ has a stationary point whenever:

$$\frac{\partial f}{\partial x} = \frac{\partial f}{\partial y} = 0$$

Let (a,b) be such a point, and let:

$$\frac{\partial^2 f}{\partial x^2} = A \qquad \frac{\partial^2 f}{\partial x \, \partial y} = H \qquad \frac{\partial^2 f}{\partial y^2} = B$$

all at that point, then:

If $H^2 - AB > 0$, $f(x,y)$ has a saddle point at (a,b)

If $H^2 - AB < 0$ and if $A < 0$, $f(x,y)$ has a maximum at (a,b), but if $A > 0$, $f(x,y)$ has a minimum at (a,b).

If $H^2 = AB$, higher derivatives need to be considered.

2.3 Integral

$$\int_a^b f(x)\, dx = \lim_{N \to \infty} \sum_{n=0}^{N-1} f\left(a + \frac{n(b-a)}{N}\right)\left(\frac{b-a}{N}\right)$$

$$= \lim_{N \to \infty} \sum_{n=1}^{N} f(a + (n-1)\, \partial x)\, \partial x$$

where $\partial x = (b-a)/N$.

If u and v are functions of x, then:

$$\int usv'\, dx = uv - \int u'v\, dx \qquad \text{(integration by parts)}$$

2.4 Derivatives and integrals

y	$\dfrac{dy}{dx}$	$\int y\, dx$
x^n	nx^{n-1}	$\dfrac{x^{n+1}}{(n+1)}$

y	$\dfrac{dy}{dx}$	$\int y\,dx$
$\dfrac{1}{x}$	$-\dfrac{1}{x^2}$	$\ln(x)$
e^{ax}	ae^{ax}	$\dfrac{e^{ax}}{a}$
$\ln(x)$	$\dfrac{1}{x}$	$x[\ln(x)-1]$
$\log_a x$	$\dfrac{1}{x}\log_a e$	$x\log_a\left(\dfrac{x}{e}\right)$
$\sin ax$	$a\cos ax$	$-\dfrac{1}{a}\cos ax$
$\cos ax$	$-a\sin ax$	$\dfrac{1}{a}\sin ax$
$\tan ax$	$a\sec^2 ax$	$-\dfrac{1}{a}\ln(\cos ax)$
$\cot ax$	$-\csc^2 ax$	$\dfrac{1}{a}\ln(\sin ax)$
$\sec ax$	$a\tan ax\sec ax$	$\dfrac{1}{a}\ln(\sec ax+\tan ax)$
$\csc ax$	$-a\cot ax\csc ax$	$\dfrac{1}{a}\ln(\csc ax-\cot ax)$
$\arcsin\left(\dfrac{x}{a}\right)$	$(a^2-x^2)^{-1/2}$	$x\arcsin\left(\dfrac{x}{a}\right)+(a^2-x^2)^{1/2}$
$\arccos\left(\dfrac{x}{a}\right)$	$-(a^2-x^2)^{-1/2}$	$x\arccos\left(\dfrac{x}{a}\right)-(a^2-x^2)^{-1/2}$
$\arctan\left(\dfrac{x}{a}\right)$	$\dfrac{a}{(a^2+x^2)}$	$x\arctan\left(\dfrac{x}{a}\right)-\tfrac{1}{2}a\ln(a^2+x^2)$

24 Derivatives and integrals

y	$\dfrac{dy}{dx}$	$\int y\, dx$
$\operatorname{arccot}\left(\dfrac{x}{a}\right)$	$\dfrac{-a}{(a^2+x^2)}$	$x \operatorname{arccot}\left(\dfrac{x}{a}\right) + \tfrac{1}{2} a \ln(a^2+x^2)$
$\operatorname{arcsec}\left(\dfrac{x}{a}\right)$	$\dfrac{a(x^2-b^2)^{-1/2}}{x}$	$x \operatorname{arcsec}\left(\dfrac{x}{a}\right) - a \ln[x+(x^2-a^2)^{1/2}]$
$\operatorname{arccosec}\left(\dfrac{x}{a}\right)$	$\dfrac{-a(x^2-a^2)^{-1/2}}{x}$	$x \operatorname{arccosec}\left(\dfrac{x}{a}\right) + a \ln[x+(x^2-a^2)^{1/2}]$
$\sinh ax$	$a \cosh ax$	$\dfrac{1}{a} \cosh ax$
$\cosh ax$	$a \sinh ax$	$\dfrac{1}{a} \sinh ax$
$\tanh ax$	$a \operatorname{sech}^2 ax$	$\dfrac{1}{a} \ln(\cosh ax)$
$\coth ax$	$-a \operatorname{cosech}^2 ax$	$\dfrac{1}{a} \ln(\sinh ax)$
$\operatorname{sech} ax$	$-a \tanh ax\ \operatorname{sechsech} ax$	$\dfrac{2}{a} \arctan(e^{ax})$
$\operatorname{cosech} ax$	$-a \coth ax\ \operatorname{cosech} ax$	$\dfrac{1}{a} \ln\left(\tanh \dfrac{ax}{2}\right)$
$\operatorname{arsinh}\left(\dfrac{x}{a}\right)$	$(x^2+a^2)^{-1/2}$	$x \operatorname{arsinh}\left(\dfrac{x}{a}\right) - (x^2+a^2)^{1/2}$
$\operatorname{arcosh}\left(\dfrac{x}{a}\right)$	$(x^2-a^2)^{-1/2}$	$x \operatorname{arcosh}\left(\dfrac{x}{a}\right) - (x^2-a^2)^{1/2}$
$\operatorname{artanh}\left(\dfrac{x}{a}\right)$	$a(a^2-x^2)^{-1}$	$x \operatorname{artanh}\left(\dfrac{x}{a}\right) + \tfrac{1}{2} a \ln(a^2-x^2)$

Calculus 25

y	$\dfrac{dy}{dx}$	$\int y\,dx$
$\operatorname{arcoth}\left(\dfrac{x}{a}\right)$	$-a(x^2-a^2)^{-1}$	$x\operatorname{arcoth}\left(\dfrac{x}{a}\right)$ $+\tfrac{1}{2}a\ln(x^2-a^2)$
$\operatorname{arsech}\left(\dfrac{x}{a}\right)$	$\dfrac{-a(a^2-x^2)^{-1/2}}{x}$	$x\operatorname{arsech}\left(\dfrac{x}{a}\right)+a\arcsin\left(\dfrac{x}{a}\right)$
$\operatorname{arcosech}\left(\dfrac{x}{a}\right)$	$\dfrac{-a(x^2+a^2)^{-1/2}}{x}$	$x\operatorname{arcosech}\left(\dfrac{x}{a}\right)$ $+a\operatorname{arsinh}\left(\dfrac{x}{a}\right)$
$(x^2\pm a^2)^{1/2}$		$\tfrac{1}{2}x(x^2\pm a^2)^{1/2}$ $\pm\tfrac{1}{2}a^2\operatorname{arsinh}\left(\dfrac{x}{a}\right)$
$(a^2-x^2)^{1/2}$		$\tfrac{1}{2}x(a^2-x^2)^{1/2}$ $+\tfrac{1}{2}a^2\operatorname{arcsinh}\left(\dfrac{x}{a}\right)$ $\tfrac{1}{2}x(a^2-x^2)^{1/2}$ $+\tfrac{1}{2}a^2\arcsin\left(\dfrac{x}{a}\right)$
$(x^2\pm a^2)^p x$		$\dfrac{\tfrac{1}{2}(x^2\pm a^2)^{p+1}}{(p+1)}\quad(p\neq -1)$ $\tfrac{1}{2}\ln(x^2\pm a^2)\quad(p=-1)$
$(a^2-x^2)^p x$		$\dfrac{-\tfrac{1}{2}(a^2-x^2)^{p+1}}{(p+1)}\quad(p\neq -1)$ $-\tfrac{1}{2}\ln(a^2-x^2)\quad(p=-1)$
$x(ax^2+b)^p$		$\dfrac{(ax^2+b)^{p+1}}{2a(p+1)}\quad(p\neq -1)$ $\dfrac{\ln(ax^2+b)}{2a}\quad(p=-1)$

26 Derivatives and integrals

y	$\dfrac{dy}{dx}$	$\int y\,dx$
$(2ax - x^2)^{-1/2}$		$\arccos\left(\dfrac{a-x}{a}\right)$
$(a^2 \sin^2 x + b^2 \cos^2 x)^{-1}$		$\dfrac{1}{ab}\arctan\left(\dfrac{a}{b}\tan x\right)$
$(a^2 \sin^2 x - b^2 \cos^2 x)^{-1}$		$-\dfrac{1}{ab}\operatorname{artanh}\left(\dfrac{a}{b}\tan x\right)$
$e^{ax}\sin bx$		$e^{ax}\dfrac{a\sin bx - b\cos bx}{a^2 + b^2}$
$e^{ax}\cos bx$		$e^{ax}\dfrac{a\cos bx + b\sin bx}{a^2 + b^2}$

y	$\int y\,dx$	
$\sin mx \sin nx$	$\dfrac{1}{2}\dfrac{\sin(m-n)x}{m-n} - \dfrac{1}{2}\dfrac{\sin(m+n)x}{m+n}$	$(m \neq n)$
	$\dfrac{1}{2}\left(x - \dfrac{\sin 2xm}{2m}\right)$	$(m = n)$
$\sin mx \cos nx$	$-\dfrac{1}{2}\dfrac{\cos(m+n)x}{m+n} - \dfrac{1}{2}\dfrac{\cos(m-n)x}{m-n}$	$(m \neq n)$
	$-\dfrac{1}{2}\dfrac{\cos 2mx}{2m}$	$(m = n)$
$\cos mx \cos nx$	$\dfrac{1}{2}\dfrac{\sin(m+n)x}{m+n} + \dfrac{1}{2}\dfrac{\sin(m-n)x}{m-n}$	$(m \neq n)$
	$\dfrac{1}{2}\left(x + \dfrac{\sin 2mx}{2m}\right)$	$(m = n)$

2.5 Standard substitutions

Integral a function of	Substitute
$a^2 - x^2$	$x = a \sin \theta$ or $x = a \cos \theta$
$a^2 + x^2$	$x = a \tan \theta$ or $x = a \sinh \theta$
$x^2 - a^2$	$x = a \sec \theta$ or $x = a \cosh \theta$

2.6 Reduction formulae

$$\int \sin^m x \, dx = -\frac{1}{m} \sin^{m-1} x \cos x + \frac{m-1}{m} \int \sin^{m-2} x \, dx$$

$$\int \cos^m x \, dx = \frac{1}{m} \cos^{m-1} x \sin x + \frac{m-1}{m} \int \cos^{m-2} x \, dx$$

$$\int \sin^m x \cos^n x \, dx = \frac{\sin^{m+1} x \cos^{n-1} x}{m+n}$$
$$+ \frac{n-1}{m+n} \int \sin^m x \cos^{n-2} x \, dx$$

If the integrand is a rational function of $\sin x$ and/or $\cos x$, substitute $t = \tan \frac{1}{2} x$, then:

$$\sin x = \frac{1}{1+t^2} \qquad \cos x = \frac{1-t^2}{1+t^2} \qquad dx = \frac{2dt}{1+t^2}$$

2.7 Numerical integration

2.7.1 Trapezoidal rule

See Figure 2.1

$$\int_{x_1}^{x_2} y \, dx = \frac{1}{2} h (y_1 + y_2) + O(h^3)$$

28 Numerical integration

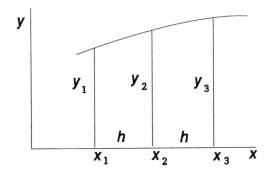

Figure 2.1 Numerical integration

2.7.2 Simpson's rule

$$\int_{x_1}^{x_2} y \, dx = \frac{2h \, (y_1 + 4y_2 + y_3)}{6} + O \, (h^5)$$

2.7.3 Change of variable in double integral

$$\iint f(x,y) \, dx \, dy = \iint F(u,v) \, |J| \, du \, dv$$

where:

$$J = \frac{\partial(x,y)}{\partial(u,v)} = \begin{vmatrix} \frac{\partial x}{\partial u} & \frac{\partial x}{\partial v} \\ \frac{\partial y}{\partial u} & \frac{\partial y}{\partial v} \end{vmatrix} = \begin{vmatrix} \frac{\partial x}{\partial u} & \frac{\partial y}{\partial u} \\ \frac{\partial x}{\partial v} & \frac{\partial y}{\partial v} \end{vmatrix}$$

is the Jacobian of the transformation.

2.7.4 Differential mean value theorem

$$\frac{f(x+h) - f(x)}{h} = f'(x + \theta h) \qquad 0 < \theta < 1$$

2.7.5 Integral mean value theorem

$$\int_a^b f(x)\, g(x)\, dx = g(a + \theta h) \int_a^b f(x)\, dx$$
$$h = b - a \qquad 0 < \theta < 1$$

2.8 Vector calculus

Let $s(x,y,z)$ be a scalar function of position and let $v(x,y,x) = iv_x(x,y,z) + jv_y(x,y,z) + kv_z(x,y,z)$ be a vector function of position. Define:

$$\nabla = i\frac{\partial}{\partial x} + j\frac{\partial}{\partial y} + k\frac{\partial}{\partial z}$$

so that:

$$\nabla \cdot \nabla = \nabla^2 = \frac{\partial^2}{\partial x^2} + \frac{\partial^2}{\partial y^2} + \frac{\partial^2}{\partial z^2}$$

then:

$$\text{grad } s = \nabla s = i\frac{\partial s}{\partial x} + j\frac{\partial s}{\partial y} + k\frac{\partial s}{\partial z}$$
$$\text{div } v = \nabla \cdot v = \frac{\partial v_x}{\partial x} + \frac{\partial v_y}{\partial y} + \frac{\partial v_z}{\partial z}$$
$$\text{curl } v = \nabla \times v = i\left(\frac{\partial v_z}{\partial y} - \frac{\partial v_y}{\partial z}\right) + j\left(\frac{\partial v_x}{\partial z} - \frac{\partial v_z}{\partial x}\right) + k\left(\frac{\partial v_y}{\partial x} - \frac{\partial v_x}{\partial y}\right)$$

The following identities are then true:

div $(sv) = s$ div $v + (\text{grad } s) \cdot v$
curl $(sv) + s$ curl $v + (\text{grad } s) \times v$
div $(u \times v) = v \cdot \text{curl } u - u \cdot \text{curl } v$
curl $(u \times v) = u$ div $v - v$ div $u + (v \cdot \nabla) u - (u \cdot \nabla) v$

Vector calculus

div grad $s = \nabla^2 s$
div curl $v = 0$
curl grad $s = 0$
curl curl v = grad (div v) $- \nabla^2 v$
where ∇^2 operates on each component of v.
$v \times$ curl $v + (v \cdot \nabla) v$ = grad $\tfrac{1}{2} v^2$

Potentials:

If curl $v = 0$, v = grad φ where φ is a scalar potential.
If div $v + 0$, v = curl A where A is a vector potential.

3. Series and transforms

3.1 Arithmetic series

Sum of n terms:

$$S_n = a + (a + d) + (a + 2d) + \ldots + [a + (n-1)d]$$
$$= \frac{n[2a + (n-1)d]}{2}$$
$$= \frac{n(a+l)}{2}$$

3.2 Geometric series

Sum of n terms:

$$S_n = a + ar + ar^2 + \ldots + ar^{n-1} = \frac{a(1-r^n)}{(1-r)} \qquad (|r| < 1)$$

$$S_\infty = \frac{a}{(1-r)}$$

3.3 Binomial series

$$(1+x)^p = 1 + px + \frac{p(p-1)}{2!}x^2 + \frac{p(p-1)(p-2)}{3!}x^3 + \ldots$$

If p is a positive integer the series terminates with the term in x^p and is valid for all x; otherwise the series does not terminate and is valid only for $-1 < x < 1$.

3.4 Taylor's series

Infinite form:

$$f(x+h) = f(x) + hf'(x) + \frac{h^2}{2!}f''(x) + \ldots + \frac{h^2}{n!}f^{(n)}(x) + \ldots$$

Finite form:

$$f(x+h) = f(x) + hf'(x) + \frac{h^2}{2!}f''(x) + \ldots$$
$$+ \frac{h^n}{n!}f^{(n)}(x) + \frac{h^{n+1}}{(n+1)!}f^{(n+1)}(x+\lambda h)$$

where $0 \leq \lambda \leq 1$.

3.5 Maclaurin's series

$$f(x) = f(0) + xf'(0) + \frac{x^2}{2!}f''(0) + \ldots + \frac{x^n}{n!}f^{(n)}(0) + \ldots$$

Neither of these series is necessarily convergent, but both usually are for appropriate ranges of values of h and of x respectively.

3.6 Laurent's series

If a function $f(z)$ of a complex variable is analytic on and everywhere between two concentric circles centre a, then at any point in this region:

$$f(z) = a_0 + a_1(z-a) + \ldots + \frac{b_1}{(z-a)} + \frac{b_2}{(z-a)^2} + \ldots$$

This series is often applicable when Taylor's series is not.

3.7 Power series for real variables

	Math	Comp
$e^x = 1 + x + \dfrac{x^2}{2!} + \ldots$	all x	$|x| \leq 1$
$\ln(1+x) = x - \dfrac{x^2}{2} + \dfrac{x^3}{3} - \dfrac{x^4}{4} + \ldots$	$-1 < x \leq 1$	
$\sin x = x - \dfrac{x^3}{3!} + \dfrac{x^5}{5!} - \dfrac{x^7}{7!} + \ldots$	all x	$|x| \leq 1$
$\cos x = 1 - \dfrac{x^2}{2!} + \dfrac{x^4}{4!} - \dfrac{x^6}{6!} + \ldots$	all x	$|x| \leq 1$
$\tan x = x + \dfrac{x^3}{3} + \dfrac{2x^5}{15} - \dfrac{17x^7}{315} + \ldots$		$|x| < \dfrac{\pi}{2}$
$\arctan x = x - \dfrac{x^3}{3} + \dfrac{x^5}{5} - \dfrac{x^7}{7} + \ldots$		$|x| \leq 1$
$\sinh x = x + \dfrac{x^3}{3!} + \dfrac{x^5}{5!} + \dfrac{x^7}{7!} + \ldots$	all x	$|x| \leq 1$
$\cosh x = 1 + \dfrac{x^2}{2!} + \dfrac{x^4}{4!} + \dfrac{x^6}{6!} + \ldots$	all x	$|x| \leq 1$

The column headed 'Math' contains the range of values of the variable x for which the series is convergent in the pure mathematical sense. In some cases a different range of values is given in the column headed 'Comp', to reduce the rounding errors which arise when computers are used.

3.8 Integer series

$$\sum_{n=1}^{N} n = 1 + 2 + 3 + 4 + \ldots + N = \frac{N(N+1)}{2}$$

$$\sum_{n=1}^{N} n^2 = 1^2 + 2^2 + 3^2 + 4^2 + \ldots + N^2 = \frac{N(N+1)(2N+1)}{6}$$

$$\sum_{n=1}^{N} n^3 = 1^3 + 2^3 + 3^3 + 4^3 + \ldots + N^3 = \frac{N^2(N+1)^2}{4}$$

$$\sum_{n=1}^{\infty} \frac{(-1)^{n+1}}{n} = 1 - \frac{1}{2} + \frac{1}{3} - \frac{1}{4} + \ldots = \ln(2) \quad \text{(see } \ln(1+x)\text{)}$$

$$\sum_{n=1}^{\infty} \frac{(-1)^{n+1}}{2n-1} = 1 - \frac{1}{3} + \frac{1}{5} - \frac{1}{7} + \ldots = \frac{\pi}{4} \quad \text{(see arctan } x\text{)}$$

$$\sum_{n=1}^{\infty} \frac{1}{n^2} = 1 + \frac{1}{4} + \frac{1}{9} + \frac{1}{16} + \ldots = \frac{\pi^2}{6}$$

$$\sum_{n=1}^{N} n(n+1)(n+2)\ldots(n+r)$$
$$= 1.2.3\ldots + 2.3.4\ldots + 3.4.5\ldots + N(N+1)(N+2)\ldots(N+r)$$
$$= \frac{N(N+1)(N+2)\ldots(N+r+1)}{r+2}$$

3.9 Fourier series

$$f(\theta) = \tfrac{1}{2} a_0 + \sum_{n=1}^{\infty} (a_n \cos n\theta + b_n \sin n\theta)$$

with:

$$a_n = \frac{1}{\pi} \int_0^{2\pi} f(\Theta) \cos n\Theta \, d\Theta$$
$$b_n = \frac{1}{\pi} \int_0^{2\pi} f(\Theta) \sin n\Theta \, d\Theta$$

or:

$$f(\theta) = \sum_{n=-\infty}^{\infty} c_n \exp(jn\theta)$$

with:

$$c_n = \frac{1}{2\pi} \int_0^{2\pi} f(\Theta) \exp(-jn\Theta)\, d\Theta = \begin{cases} \frac{1}{2}(a_n + jb_n) & n < 0 \\ \frac{1}{2}(a_n - jb_n) & n > 0 \end{cases}$$

The above expressions for Fourier series are valid for functions having at most a finite number of discontinuities within the period 0 to 2 of the variable of integration.

3.10 Rectified sine wave

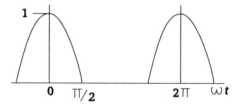

Figure 3.1 Half wave

$$f(\omega t) = \frac{1}{\pi} + \frac{1}{2}\cos\omega t + \frac{2}{\pi}\sum_{n=1}^{\infty}(-1)^{n+1}\frac{\cos 2n\omega t}{4n^2 - 1}$$

Figure 3.2 *p-phase*

$$f(\omega t) = \frac{\sin(\pi/p)}{\pi/p} + \frac{2p}{\pi}\sin\left(\frac{\pi}{p}\right)\sum_{n=1}^{\infty}(-1)^{n+1}\frac{\cos np\omega t}{p^2 n^2 - 1}$$

3.11 Square wave

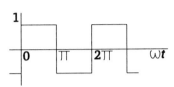

Figure 3.3 Square wave

$$f(\omega t) = \frac{4}{\pi}\sum_{n=1}^{\infty}\frac{\sin(2n-1)\omega t}{(2n-1)}$$

3.12 Triangle wave

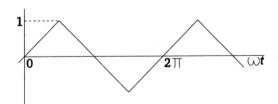

Figure 3.4 Triangular wave

$$f(\omega t) = \frac{8}{\pi^2}\sum_{n+1}^{\infty}(-1)^{n+1}\frac{\sin(2n-1)\omega t}{(2n-1)^2}$$

Series and transforms 37

3.13 Sawtooth wave

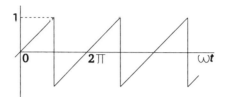

Figure 3.5 Sawtooth wave

$$f(\omega t) = \frac{2}{\pi} \sum_{n=1}^{\infty} (-1)^{n+1} \frac{\sin n\omega t}{n}$$

3.14 Pulse wave

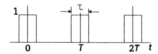

Figure 3.6 Pulse wave

$$f(t) = \frac{\tau}{T} + \frac{2\tau}{T} \sum_{n=1}^{\infty} \frac{\sin(n\omega\tau/T)}{n\pi\tau/T} \cos\left(\frac{2n\pi t}{T}\right)$$

3.15 Fourier transforms

Among other applications, these are used for converting from the time domain to the frequency domain.

Basic formulae:

38 Fourier transforms

$$\int_{-\infty}^{\infty} U(f) \exp(j2\pi ft) \, df = u(t) \leftrightarrow U(f) = \int_{-\infty}^{\infty} u(t) \exp(-j2\pi ft) \, dt$$

Change of sign and complex conjugates:

$$u(-t) \leftrightarrow U(-f) \qquad u^*(t) \leftrightarrow U^*(-f)$$

Time and frequency shifts (τ and φ constant):

$$u(t-\tau) \leftrightarrow U(f) \exp(-2\pi f\tau) \quad \exp(j2\pi\varphi t) \, u(t) \leftrightarrow U(f-\varphi)$$

Scaling (T constant):

$$u\left(\frac{t}{T}\right) \leftrightarrow TU(fT)$$

Products and convolutions:

$$u(t) \dagger v(t) \leftrightarrow U(f) \, V(f)$$
$$u(t) \, v(t) \leftrightarrow U(f) \dagger V(f)$$

Differentiation:

$$u'(t) \leftrightarrow j2\pi f U(f)$$
$$-j2\pi t u(t) \leftrightarrow U'(f)$$
$$\frac{\partial u(t, \alpha)}{\partial \alpha} \leftrightarrow \frac{\partial (U-f, \alpha)}{\partial \alpha}$$

Integration ($U(0) = 0$, a and b real and constants):

$$\int_{-\alpha}^{t} u(\tau) \, d\tau \leftrightarrow \frac{U(f)}{j2\pi f}$$
$$\int_{a}^{b} v(t, \alpha) \, d\alpha \leftrightarrow \int_{a}^{b} V(f, \alpha) \, d\alpha$$

Interchange of functions:

$U(t) \leftrightarrow u(-f)$

Dirac delta functions:

$\partial(t) \leftrightarrow 1 \qquad \exp(j2\pi f_0 t) \leftrightarrow \partial(f - f_0)$

Unit length, unit amplitude pulse, centred on $t = 0$:

$\text{rect}(t) \leftrightarrow \dfrac{\sin \pi f}{\pi f}$

Gaussian distribution:

$\exp(-\pi t^2) \leftrightarrow \exp(-\pi f^2)$

Repeated and impulse (delta function) sampled waveforms:

$$\sum_{-\infty}^{\infty} u(t - nT) \leftrightarrow \left(\frac{1}{T}\right) U(f) \sum_{-\infty}^{\infty} \partial\left(f - \frac{n}{T}\right)$$

$$u(t) \sum_{-\infty}^{\infty} \partial(t - nT) \leftrightarrow \left(\frac{1}{T}\right) \sum_{-\infty}^{\infty} U\left(f - \frac{n}{T}\right)$$

Parseval's lemma:

$$\int_{-\infty}^{\infty} u(t)\, v^*(t)\, dt = \int_{-\infty}^{\infty} U(f)\, V^*(f)\, df$$

$$\int_{-\infty}^{\infty} |u(t)|^2\, dt = \int_{-\infty}^{\infty} |U(f)|^2\, df$$

3.16 Laplace transforms

$$\bar{x}_x = \int_0^{\infty} x(t) \exp(-st)$$

Laplace transforms

Function	Transform	Remarks
$e^{-\alpha t}$	$\dfrac{1}{s+\alpha}$	
$\sin \omega t$	$\dfrac{\omega}{s^2+\omega^2}$	
$\cos \omega t$	$\dfrac{s}{s^2+\omega^2}$	
$\sinh \omega t$	$\dfrac{\omega}{s^2-\omega^2}$	
$\cosh \omega t$	$\dfrac{s}{s^2}-\omega^2$	
t^n	$\dfrac{n!}{s^{n+1}}$	
1	$\dfrac{1}{s}$	
$H(t-\tau)$	$\dfrac{1}{s}\exp(-s\tau)$	Heaviside step function
$x(t-\tau)\,H(t-\tau)$	$\exp(-s\tau)\,\bar{x}(s)$	Shift in t
$\partial(t-\tau)$	$\exp(-s\tau)$	Dirac delta function
$\exp(-\alpha t)\,x(t)$	$\bar{x}(s+\alpha)$	Shift in s
$\exp(-\alpha t)\sin \omega t$	$\dfrac{\omega}{(s+\alpha)^2+\omega^2}$	
$\exp(-\alpha t)\cos \omega t$	$\dfrac{(s+\alpha)}{(s+\alpha)^2+\omega^2}$	
$t\,x(t)$	$-\dfrac{d\bar{x}(s)}{ds}$	
$\dfrac{dx(t)}{dt}=x'(t)$	$s\bar{x}(s)-x(0)$	
$\dfrac{d^2x(t)}{dt^2}=x''(t)$	$x^2\bar{x}(s)-sx(0)-x'(0)$	

Function	Transform	Remarks
$\dfrac{d^n x(t)}{dx^n} = x^{(n)}(t)$	$s^n \bar{x}(s) - s^{n-1} x(0)$ $- s^{n-2} x'(0) \ldots$ $- s x^{(n-2)}(0) - x^{(n-1)}(0)$	

Convolution integral

$$\int_0^t x_1(\sigma) x_2(t - \sigma)\, d\sigma \to \bar{x}_1(s)\, \bar{x}_2(s)$$

4. Matrices and determinants

4.1 Linear simultaneous equations

This set of equations:

$$a_{11}x_1 + a_{12}x_2 + \ldots + a_{1n}x_n = b_1$$
$$a_{21}x_1 + a_{22}x_2 + \ldots + a_{2n}x_n = b_2$$
$$\ldots$$
$$a_{n1}x_1 + a_{n2}x_2 + \ldots + a_{nn}x_n = b_n$$

may be written symbolically:

$$Ax = b$$

in which A is the *matrix* of the coefficients a_{ij} and x and b are the column matrices or *vectors* $(x_1 \ldots x_n)$ and $(b_1 \ldots b_n)$. In this case the matrix A is square $(n \times n)$. The equations can be solved unless two or more of them are not independent, in which case:

$$\det A = |A| = 0$$

and there then exists non-zero solutions x_i only if $b = 0$. If $\det A \neq 0$, there exist non-zero solutions only if $b \neq 0$. When $\det A = 0$, A is *singular*.

4.2 Matrix arithmetic

If A and B are both matrices of m rows and n columns they are *conformable* and:

Matrices and determinants 43

$A \pm B = C$ where $C_{ij} = A_{ij} \pm B_{ij}$

4.2.1 Product

If A is an $m \times n$ matrix and B an $n \times l$, the product AB is defined by:

$$(AB)_{ij} = \sum_{k=1}^{n} (A)_{ik} (B)_{kj}$$

In this case, if $l \neq m$, the product BA will not exist.

4.2.2 Transpose

The transpose of A is written A' or A^t and is the matrix whose rows are the columns of A, i.e.:

$(A')_{ij} = (A)_{ji}$

A square matrix may be equal to its transpose and it is then said to be *symmetrical*. If the product AB exists, then:

$(AB)' = B'A'$

4.2.3 Adjoint

The *adjoint* of a square matrix A is defined as B, where:

$(B)_{ij} = (A)_{ji}$

and A_{ji} is the *cofactor* of a_{ji} in det A.

4.2.4 Inverse

If A is non-singular, the *inverse* $A^{-1} = 1$ is given by:

$$A^{-1} = \frac{\text{adj } A}{\det A}$$

$$A^{-1}A = AA^{-1} = I$$

the *unit* matrix.

$$(AB)^{-1} = B^{-1}A^{-1}$$

if both inverses exist. The original equations $Ax = b$ have the solutions $x = A^{-1}b$ if the inverse exists.

4.2.5 Orthogonality

A matrix A is orthogonal if $AA^t = 1$. If A is the matrix of a coordinate transformation $X = AY$ from variables y_i to variables x_i, then if A is orthogonal $X^tX = Y^tY$, or:

$$\sum_{i=1}^{n} x_i^2 = \sum_{i=1}^{n} y_i^2$$

4.2.6 Eigenvalues and eigenvectors

The equation:

$$Ax = \lambda x$$

where A is a square matrix, x a column vector and λ a number (in general complex) has at most n solutions (x, λ). The values of λ are *eigenvalues* and those of x *eigenvectors* of the matrix A. The relation may be written:

$$(A - \lambda I)x = 0$$

so that if $x \neq 0$, the equation $A - \lambda I = 0$ gives the eigenvalues. If A is symmetric and real, the eigenvalues are real. If A is symmetric, the eigenvectors are orthogonal. If A is not symmetric, the eigenvalues are complex and the eigenvectors are not orthogonal.

4.3 Coordinate transformation

Suppose x and y are two vectors related by the equation:

$$y = Ax$$

when their components are expressed in one orthogonal system and that a second orthogonal system has unit vectors $u_1, u_2, ..., u_n$ expressed in the first system. The components of x and y expressed in the new system will be x' and y', where:

$$x' = U^t x \qquad\qquad y' = U^t y$$

and U^t is the orthogonal matrix whose rows are the unit vectors u_1^t, u_2^t, etc. Then:

$$y' = U^t y = U^t A x = U^t A U x'$$

$$y' = A' x'$$

where:

$$A' = U^t A U$$

Matrices A and A' are *congruent*.

4.4 Determinants

The determinant:

$$D = \begin{vmatrix} a_{11} & a_{12} & \cdots & a_{1n} \\ a_{21} & a_{22} & \cdots & a_{2n} \\ \cdot & \cdot & \cdot & \cdot \\ \cdot & \cdot & \cdot & \cdot \\ \cdot & \cdot & \cdot & \cdot \\ a_{n1} & a_{n2} & \cdots & a_{nn} \end{vmatrix}$$

is defined as follows. The first suffix in a_{rs} refers to the row, the second to the column which contains a_{rs}. Denote by M_{rs} the determinant left by deleting the rth row and the sth column from D, then:

$$D = \sum_{k=1}^{n} (-1)^{k+1} a_{1k} M_{1k}$$

gives the value of D in terms of determinants of order $n-1$, by repeated application of the determinant in terms of the elements a_{rs}.

4.5 Properties of determinants

If the rows of $\left| a_{rs} \right|$ are identical with the columns of $\left| b_{sr} \right|$, $a_{rs} = b_{sr}$ and:

$$\left| a_{rs} \right| = \left| b_{sr} \right|$$

that is, the *transposed* determinant is equal to the original.

If two rows or two columns are interchanged, the numerical value of the determinant is unaltered, but the sign will be changed if the permutation of rows or columns is odd.

If two rows or two columns are identical, the determinant is zero.

If any row or column is zero, so is the determinant.

If each element of the pth row or column of the determinant of c_{rs} is equal to the sum of the elements of the same row or column indeterminants of a_{rs} and b_{rs} then:

$$|c_{rs}| = |a_{rs}| + |b_{rs}|$$

The addition of any multiple of one row (or column) to another row (or column) does not alter the value of the determinant.

4.5.1 Minor

If row p and column q are deleted from $|a_{rs}|$ the remaining determinant of M_{pq} is called the *minor* of a_{pq}.

4.5.2 Cofactor

The *cofactor* of a_{pq} is the minor of a_{pq} prefixed by the sign which the product $M_{pq} a_{pq}$ would have in the expansion of the determinant and is denoted by A_{pq}:

$$A_{pq} = (-1)^{p+q} M_{pq}$$

A determinant a_{ij} in which $a_{ij} = a_{ji}$ for all i and j is called *symmetric*, whilst if $a_{ij} = -a_{ji}$ for all i and j the determinant is *skew-symmetric*. It follows that $a_{ii} = 0$ for all i in a skew-symmetric determinant.

4.6 Numeric solution of linear equations

Evaluation of a determinant by direct expansion in terms of elements and cofactors is disastrously slow and other methods are available, usually programmed on any existing computer system.

The system of equations may be written:

48 Numeric solution of linear equations

$$\begin{bmatrix} a_{11} & a_{12} & \ldots & a_{1n} \\ a_{21} & a_{22} & \ldots & a_{2n} \\ . & . & . & . \\ . & . & . & . \\ . & . & . & . \\ a_{n1} & a_{n2} & \ldots & a_{nn} \end{bmatrix} x_1 \begin{bmatrix} x_1 \\ x_2 \\ . \\ . \\ . \\ x_n \end{bmatrix} = \begin{bmatrix} b_1 \\ b_2 \\ . \\ . \\ . \\ b_n \end{bmatrix}$$

the variable x_1 is eliminated from the last $n-1$ equations by adding a multiple $-a_{i1}/a_{11}$ of the first row to the ith, obtaining:

$$\begin{bmatrix} a_{11} & a_{12} & \ldots & a_{1n} \\ 0 & a'_{22} & \ldots & a'_{2n} \\ . & . & . & . \\ . & . & . & . \\ . & . & . & . \\ 0 & 0 & \ldots & a''_{nn} \end{bmatrix} x_1 \begin{bmatrix} x_1 \\ x_2 \\ . \\ . \\ . \\ x_n \end{bmatrix} = \begin{bmatrix} b_1 \\ b_2 \\ . \\ . \\ . \\ b''_n \end{bmatrix}$$

where primes indicate altered coefficients. This process may be continued by eliminating x_2 from rows 3 to n, and so on

Alternatively the process may be applied to the system of equations in the form:

$$AX = Ib$$

where I is the unit matrix and the same operations carried out upon I as upon A. If the process is continued after reaching the upper triangular form, the matrix A can eventually be reduced to diagonal form. Finally, each equation is divided by the corresponding diagonal element of A, thus reducing A to the unit matrix. The system is now in the form:

$$Ix = Bb$$

and evidently $B = A^{-1}$. The total number of operations required is $O(n^3)$.

5. Statistical analysis

5.1 Introduction

Data are available in vast quantities in all areas of telecommunications. This chapter describes the more commonly used techniques for presenting and manipulating data to obtain meaningful results.

5.2 Data presentation

Probably the most common method used to present data is by tables and graphs. For impact, or to convey information quickly, pictograms and bar charts may be used. Pie charts are useful in showing the different proportions of a unit.

A strata graph shows how the total is split amongst its constituents. For example, Figure 5.1 shows that the total revenue obtained by a PTO (public telephone operator) steadily increases with time, but that only services B and D have growth, whilst service C is reducing and may eventually become unprofitable.

Logarithmic or ratio graphs are used when one is more interested in the change in the ratios of numbers than in their absolute value. In the logarithmic graph, equal ratios are represented by equal distances.

Frequency distributions are conveniently represented by a histogram as in Figure 5.2. This shows the number of people using a given service, banded according to age group. There are very few users below 20 years and above 55 years, the most popular ages being 35 to 40 years. This information will allow the service provider to target its advertising more effectively.

In a histogram, the areas of the rectangles represent the frequencies in the different groups. Ogives, illustrated in Figure 5.3, show the cumulative frequency occurrences above or below a given value.

50 Data presentation

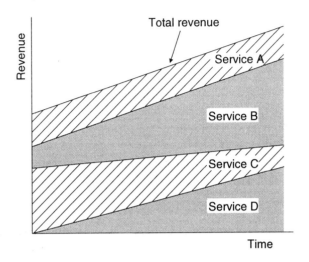

Figure 5.1 Illustration of a strata graph

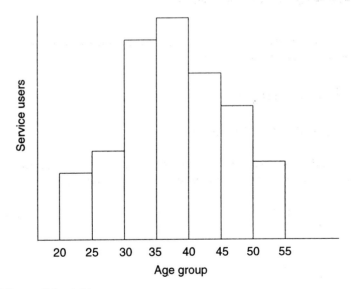

Figure 5.2 A histogram

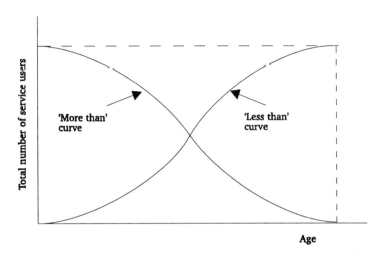

Figure 5.3 Illustration of ogives

From this curve it is possible to read off the total number of users above or below a specific age.

5.3 Averages

5.3.1 Arithmetic mean

The arithmetic mean of n numbers x_1, x_2, x_3, ..., x_n is given by Equation 5.1 which can also be written as in Equation 5.2.

$$\bar{x} = \frac{x_1 + x_2 + x_3 + \ldots + x_n}{n} \qquad (5.1)$$

$$\bar{x} = \frac{\sum_{r=1}^{n} x_r}{n} \qquad (5.2)$$

The arithmetic mean is easy to calculate and it takes into account all the figures. Its disadvantages are that it is influenced unduly by extreme values and the final result may not be a whole number, which can be absurd at times, e.g. a mean of 2.5 men.

5.3.2 Median and mode

The median or 'middle one' is found by placing all the figures in order and choosing the one in the middle or, if there are an even number of items, the mean of the two central numbers. This is a useful technique for finding the average of items which cannot be expressed in figures, e.g. shades of a colour. It is also not influenced by extreme values. However, the median is not necessarily representative of all the values.

The mode is the most 'fashionable' item, that is, the one which appears the most frequently.

5.3.3 Geometric mean

The geometric mean of n numbers $x_1, x_2, x_3, ..., x_n$ is given by Equation 5.3.

$$x_g = \left(x_1 \times x_2 \times x_3 \times ... \times x_n \right)^{1/n} \tag{5.3}$$

This technique is used to find the average of quantities which follow a geometric progression or exponential law, such as rates of change. Its advantage is that it takes into account all the numbers, but is not unduly influenced by extreme values.

5.3.4 Harmonic mean

The harmonic mean of *n* numbers $x_1, x_2, x_3, ..., x_n$ is given by Equation 5.4.

$$x_h = \frac{n}{\sum_{r=1}^{n} \frac{1}{x_r}} \tag{5.4}$$

This averaging method is used when dealing with rates or speeds or prices. As a rule when considering items expressed as A per B, if the figures are for equal As then the harmonic mean is used, but if they are for equal Bs then the arithmetic mean is used. So if a car travels over three equal distances at speeds of 5m/s, 10m/s and 15m/s the mean speed is given by the harmonic mean as in expression 5.5.

$$\frac{3}{\frac{1}{5} + \frac{1}{10} + \frac{1}{15}} = 8.18\text{m/s} \tag{5.5}$$

If, however, the car were to travel for three equal times, of say, 20 seconds at speeds of 5m/s, 10m/s and 15m/s, then the mean speed would be given by the arithmetic mean as in expression 5.6.

$$\frac{5 + 10 + 15}{3} = 10\text{m/s} \tag{5.6}$$

5.4 Dispersion from the average

5.4.1 Range and quartiles

The average represents the central figure of a series of numbers or items. It does not give any indication of the spread of the figures in the series from the average. For example measurements of errors made on two circuits A and B may result in the curves shown in Figure 5.4. Both circuits have the same calculated average errors, but circuit B has a wider deviation from the average than circuit A and at the top end its errors may be unacceptably high.

There are several ways of stating by how much the individual numbers in the series differ from the average. The range is the

54 Dispersion from the average

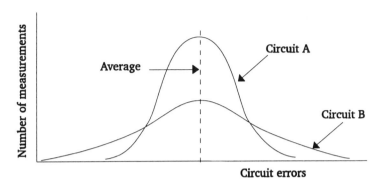

Figure 5.4 Illustration of deviation from the average

difference between the smallest and largest values. The series can also be divided into four quartiles and the dispersion stated as the inter-quartile range, which is the difference between the first and third quartile numbers, or the quartile deviation which is half this value.

The quartile deviation is easy to use and is not influenced by extreme values. However, it gives no indication of distribution between quartiles and covers only half the values in a series.

5.4.2 Mean deviation

This is found by taking the mean of the differences between each individual number in the series and the arithmetic mean, or median, of the series. Negative signs are ignored.

For a series of n numbers $x_1, x_2, x_3, ..., x_n$ having an arithmetic mean of $\bar{x}$ the mean deviation of the series is given by Equation 5.7.

$$\frac{\sum_{r=1}^{n} |x_r - \bar{x}|}{n} \tag{5.7}$$

The mean deviation takes into account all the items in the series but it is not very suitable since it ignores signs.

5.4.3 Standard deviation

This is the most common measure of dispersion. For this the arithmetic mean must be used and not the median. It is calculated by squaring deviations from the mean, so eliminating their sign, adding the numbers together, taking their mean and then the square root of the mean. Therefore, for the series of n numbers as above, the standard deviation is given by Equation 5.8.

$$\sigma = \left(\frac{\sum_{r+1}^{n} (x_r - \bar{x})^2}{n} \right)^{1/2} \quad (5.8)$$

The unit of the standard deviation is that of the original series. So if the tariff charged for a given service by different PTOs is in dollars, then the mean and the standard deviation are in dollars.

To compare two series which have different units, such as the cost of a service and the quality of that service, the coefficient of variation is used, which is unitless, as in Equation 5.9.

$$\text{Coefficient of variation} = \frac{\sigma}{\bar{x}} \times 100 \quad (5.9)$$

5.5 Skewness

The distribution shown in Figure 5.4 is symmetrical since the mean, median and mode all coincide. Figure 5.5 shows a skewed distribution with positive skewness. If the distribution bulges the other way, the skewness is said to be negative.

There are several mathematical ways of expressing skewness. They all give a measure of the deviation between the mean, median and mode and they are usually stated in relative terms for ease of comparison between series of different units. The Pearson coefficient of skewness is given by Equation 5.10.

56 Combinations and permutations

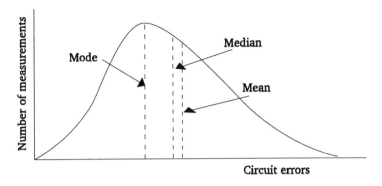

Figure 5.5 Illustration of skewness

$$P_k = \frac{\text{mean} - \text{mode}}{\text{standard deviation}} \quad (5.10)$$

Since the mode is sometimes difficult to measure this can also be stated as in Equation 5.11.

$$P_k = \frac{3(\text{mean} - \text{median})}{\text{standard deviation}} \quad (5.11)$$

5.6 Combinations and permutations

5.6.1 Combinations

Combinations are the number of ways in which a proportion can be chosen from a group. For example the number of ways in which two letters can be chosen from a group of four letters A, B, C, D is equal to 6, i.e. AB, AC, AD, BC, BD, CD. This is written as in expression 5.12.

$$^4C_2 = 6 \quad (5.12)$$

The factorial expansion is frequently used in combination calculations, where factorial n is written as in expression 5.13.

$$n! = n \times (n-1) \times (n-2) \times \ldots \times 3 \times 2 \times 1 \tag{5.13}$$

Using this, the number of combinations of r items available from a group of n is given by Equation 5.14.

$$^nC_r = \frac{n!}{r!(n-r)!} \tag{5.14}$$

5.6.2 Permutations

Combinations do not indicate any sequencing. When sequencing within each combination is involved, the result is known as a permutation. Therefore the number of permutations of two letters from four letters A, B, C, D is 12, i.e. AB, BA, AC, CA, AD, DA, BC, CB, BD, DB, CD, DC. The number of permutations of r items from a group of n is given by Equation 5.15.

$$^nP_r = \frac{n!}{(n-r)!} \tag{5.15}$$

5.7 Regression and correlation

5.7.1 Regression

Regression is a method for establishing a mathematical relationship between two variables. Several equations may be used to determine this relationship, the most common being that of a straight line. Figure 5.6 shows the number of defective public telephones which were reported at seven instances in time. This is called a scatter diagram. The points can be seen to lie approximately on the straight line AB.

58 Regression and correlation

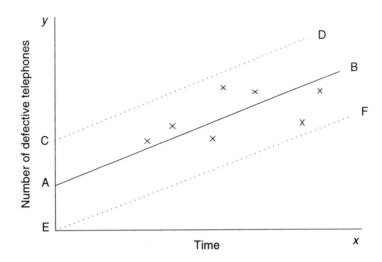

Figure 5.6 A scatter diagram

The equation of a straight line is given by Equation 5.16, where x is the independent variable, y the dependent variable, m the slope of the line and c its intercept on the y axis.

$$y = mx + c \qquad (5.16)$$

c is negative if the line intercepts the y axis on its negative part and m is negative if the line slopes the other way to that shown in Figure 5.6.

The best straight line to fit a set of points is found by the method of least squares as in Equations 5.17 and 5.18, where n is the number of points. The line passes through the mean values of x and y, i.e. $\overline{x}$ and $\overline{y}$.

$$m = \frac{\sum xy - \dfrac{\sum x \sum y}{n}}{\sum x^2 - \dfrac{(\sum x)^2}{n}} \tag{5.17}$$

$$c = \frac{\sum x \sum xy - \sum y \sum x^2}{\left(\sum x\right)^2 - n \sum x^2} \tag{5.18}$$

5.7.2 Correlation

Correlation is a technique for establishing the strength of the relationship between variables. In Figure 5.6 the individual figures are scattered on either side of a straight line and although one can approximate them by a straight line it may be required to establish if there is correlation between the readings on the x and y axes.

Several correlation coefficients exist. The product moment correlation coefficient (r) is given by Equation 5.19 or 5.20.

$$r = \frac{\sum (x - \bar{x})(y - \bar{y})}{n \, \sigma_x \sigma_y} \tag{5.19}$$

$$r = \frac{\sum (x - \bar{x})(y - \bar{y})}{\left(\sum (x - \bar{x})^2 \sum (y - \bar{y})^2 \right)^{1/2}} \tag{5.20}$$

The value of r varies from +1, when all the points lie on a straight line and the number of defects increases with time, to −1, when all the points lie on a straight line but defects decrease with time. When $r = 0$ the points are widely scattered and there is said to be no correlation between the x and y values.

The standard error of estimation in r is given by Equation 5.21.

$$S_y = \sigma_y (1 - r^2)^{1/2} \tag{5.21}$$

In about 95% of cases, the actual values will lie within plus or minus twice the standard error of estimated values given by the regression equation. This is shown by lines CD and EF in Figure 5.6. Almost all the values will be within plus or minus three times the standard error of estimated values.

It should be noted that σ_y is the variability of the y values, whereas S_y is a measure of the variability of the y values as they differ from the regression which exists between x and y. If there is no regression then $r = 0$ and $\sigma_y = S_y$.

It is often necessary to draw conclusions from the order in which items are ranked. For example, two customers may rank the styling of a telephone and we need to know if there is any correlation between their rankings. This may be done by using the rank correlation coefficient (R) given by Equation 5.22, where d is the difference between the two ranks for each item and n is the number of items.

$$R = 1 - \frac{6 \sum d^2}{n^3 - n} \tag{5.22}$$

The value of R will vary from +1 when the two ranks are identical, to −1 when they are exactly reversed.

5.8 Probability

If an event A occurs n times out of a total of m cases then the probability of occurrence is stated to be as in Equation 5.23.

$$P(A) = \frac{n}{m} \tag{5.23}$$

Probability varies between 0 and 1. If $P(A)$ is the probability of occurrence then $1 - P(A)$ is the probability that event A will not occur and it can be written as $P(\overline{A})$.

If A and B are two events then the probability that either may occur is given by Equation 5.24.

$$P(A \text{ or } B) = P(A) + P(B) - P(A \text{ and } B) \tag{5.24}$$

A special case of this probability law is when events are mutually exclusive, i.e. the occurrence of one event prevents the other from happening. Then Equation 5.25 is obtained.

$$P(A \text{ or } B) = P(A) + P(B) \tag{5.25}$$

If A and B are two events then the probability that they may occur together is given by Equation 5.26 or 5.27.

$$P(A \text{ and } B) = P(A) \times P(B|A) \tag{5.26}$$

$$P(A \text{ and } B) = P(B) \times P(A|B) \tag{5.27}$$

$P(B|A)$ is the probability that event B will occur assuming that event A has already occurred and $P(A|B)$ is the probability that event A will occur assuming that event B has already occurred. A special case of this probability law is when A and B are independent events, i.e. the occurrence of one event has no influence on the probability of the other event occurring. Then Equation 5.28 is obtained.

$$P(A \text{ and } B) = P(A) \times P(B) \tag{5.28}$$

Bayes' theorem on probability may be stated as in Equation 5.29.

$$P(A|B) = \frac{P(A)P(B|A)}{P(A)P(B|A) + P(\overline{A})P(B|\overline{A})} \tag{5.29}$$

As an example of the use of Bayes' theorem, suppose that a company discovers that 80% of those who bought its multiplexers in a year had been on the company's training course. 30% of those who bought a competitor's multiplexers had also been on the same training course. During that year the company had 20% of the multiplexer market share. The company wishes to know what percentage of buyers actually went on its training course, in order to discover the effectiveness of this course.

If B denotes that a person bought the company's product and T that he went on the training course then the problem is to find $P(B|T)$. From the data $P(B) = 0.2$, $P(\overline{B}) = 0.8$. Then from Equation 5.29 expression 5.30 is obtained.

$$P(B|T) = \frac{0.2 \times 0.8}{0.2 \times 0.8 + 0.8 \times 0.3} = 0.4 \tag{5.30}$$

5.9 Probability distributions

There are several mathematical formulae with well defined characteristics and these are known as probability distributions. If a problem can be made to fit one of these distributions then its solution is simplified. Distributions can be discrete, when the characteristic can only take certain specific values, such as 0, 1, 2, etc., or they can be continuous, where the characteristic can take any value.

5.9.1 Binomial distribution

The binomial probability distribution is given by Equation 5.31.

$$\begin{aligned}(p+q)^n &= q^n + {}^nC_1 p\, q^{n-1} + {}^nC_2 p^2 q^{n-2} \\ &\quad + \ldots + {}^nC_x p^x q^{n-x} + \ldots + p^n\end{aligned} \tag{5.31}$$

p is the probability of an event occurring, $q\ (=1-p)$ is the probability of an event not occurring and n is the number of selections.

The probability of an event occurring m successive times is given by the binomial distribution as in Equation 5.32.

$$p(m) = {}^nC_m p^m q^{n-m} \tag{5.32}$$

The binomial distribution is used for discrete events and is applicable if the probability of occurrence p of an event is constant on each trial. The mean of the distribution $B(M)$ and the standard deviation $B(S)$ are given by Equations 5.33 and 5.34.

$$B(M) = np \tag{5.33}$$

$$B(S) = (npq)^{1/2} \tag{5.34}$$

5.9.2 Poisson distribution

The Poisson distribution is used for discrete events and, like the binomial distribution, it applies to mutually independent events. It is used in cases where p and q cannot both be defined. For example, one can state the number of times a telephone circuit failed over a given period of time, but not the number of times when it did not fail.

The Poisson distribution may be considered to be the limiting case of the binomial when n is large and p is small. The probability of an event occurring m successive times is given by the Poisson distribution as in Equation 5.35.

$$p(m) = (np)^m \frac{e^{-np}}{m!} \tag{5.35}$$

The mean $P(M)$ and standard deviation $P(S)$ of the Poisson distribution are given by Equations 5.36 and 5.37.

$$P(M) = np \tag{5.36}$$

$$P(S) = (np)^{1/2} \tag{5.37}$$

Poisson probability calculations can be done by the use of probability charts as shown in Figure 5.7. This shows the probability that an event will occur at least m times when the mean (or expected) value np is known.

64 Probability distributions

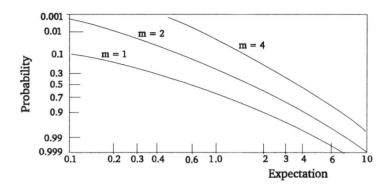

Figure 5.7 Poisson probability paper

5.9.3 Normal distribution

The normal distribution represents continuous events and is shown plotted in Figure 5.8. The x axis gives the event (for example telephone line failure) and the y axis the probability of the event occurring. The curve shows that most of the events occur close to the mean value and this is usually the case in nature. The normal curve is given by Equation 5.38, where $\bar{x}$ is the mean of the values making up the curve and σ is their standard deviation.

$$y = \frac{1}{\sigma(2\pi)^{1/2}} \exp\left(\frac{-(x-\bar{x})^2}{2\sigma^2} \right) \tag{5.38}$$

Different distributions will have different means and standard deviations but if they are distributed normally their curves will all follow Equation 5.38. These distributions can be normalised to a standard form by moving the origin of their normal curve to their mean value, shown as A in Figure 5.8. The deviation from the mean is now represented on a new scale of units given by Equation 5.39.

Statistical analysis 65

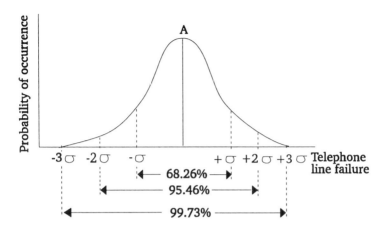

Figure 5.8 The normal curve

$$\omega = \frac{x - \bar{x}}{\sigma} \tag{5.39}$$

The standardised normal curve now becomes as in Equation 5.40.

$$y = \frac{1}{(2\pi)^{1/2}} \exp\left(\frac{-\omega^2}{2}\right) \tag{5.40}$$

The total area under the standardised normal curve is unity and the area between any two values of ω is the probability of an item from the distribution falling between these values. The normal curve extends infinitely in either direction but 68.26% of its values (area) fall between $\pm \sigma$, 95.46% between $\pm 2\sigma$, 99.73% between $\pm 3\sigma$ and 99.994% between $\pm 4\sigma$.

Table 5.1 gives the area under the normal curve for different values of ω. Since the normal curve is symmetrical, the area from $+\omega$ to $+\infty$ is the same as from $-\omega$ to $-\infty$. As an example of the use of this table, suppose that 5000 telephones have been installed in a city and that they have a mean life of 1000 weeks with a standard deviation of 100 weeks. How many telephones will fail in the first 800 weeks? From Equation 5.39, expression 5.41 is obtained.

$$\omega = \frac{(800 - 1000)}{100} = -2 \tag{5.41}$$

Ignoring the negative sign, Table 5.1 gives the probability of telephones not failing as 0.977 so that the probability of failure is 1 − 0.977 or 0.023. Therefore 5000 × 0.023 or 115 telephones are expected to fail after 800 weeks.

Table 5.1 Area under the normal curve from −∞ to ω (continued on next page)

ω	0.00	0.02	0.04	0.06	0.08
0.0	0.500	0.508	0.516	0.524	0.532
0.1	0.540	0.548	0.556	0.564	0.571
0.2	0.579	0.587	0.595	0.603	0.610
0.3	0.618	0.626	0.633	0.641	0.648
0.4	0.655	0.663	0.670	0.677	0.684
0.5	0.692	0.699	0.705	0.712	0.719
0.6	0.726	0.732	0.739	0.745	0.752
0.7	0.758	0.764	0.770	0.776	0.782
0.8	0.788	0.794	0.800	0.805	0.811
0.9	0.816	0.821	0.826	0.832	0.837
1.0	0.841	0.846	0.851	0.855	0.860
1.1	0.864	0.869	0.873	0.877	0.881
1.2	0.885	0.889	0.893	0.896	0.900
1.3	0.903	0.907	0.910	0.913	0.916
1.4	0.919	0.922	0.925	0.928	0.931
1.5	0.933	0.936	0.938	0.941	0.943
1.6	0.945	0.947	0.950	0.952	0.954
1.7	0.955	0.957	0.959	0.961	0.963

Table 5.1 Area under the normal curve from $-\infty$ to ω (continued from previous page)

1.8	0.964	0.966	0.967	0.969	0.970
1.9	0.971	0.973	0.974	0.975	0.976
2.0	0.977	0.978	0.979	0.980	0.981
2.1	0.982	0.983	0.984	0.985	0.985
2.2	0.986	0.987	0.988	0.988	0.989
2.3	0.989	0.990	0.990	0.991	0.991
2.4	0.992	0.992	0.993	0.993	0.993
2.5	0.994	0.994	0.994	0.995	0.995
2.6	0.995	0.996	0.996	0.996	0.996
2.7	0.997	0.997	0.997	0.997	0.997
2.8	0.997	0.998	0.998	0.998	0.998
2.9	0.998	0.998	0.998	0.998	0.999
3.0	0.999	0.999	0.999	0.999	0.999

5.9.4 Exponential distribution

The exponential probability distribution is a continuous distribution and is shown in Figure 5.9. It has the form given in Equation 5.42, where $\bar{x}$ is the mean of the distribution.

$$y = \frac{1}{x} \exp\left(\frac{-x}{\bar{x}}\right) \tag{5.42}$$

Whereas in the normal distribution the mean value divides the population in half, for the exponential distribution 36.8% of the population is above the average and 63.2% below the average. Table 5.2 shows the area under the exponential curve for different values of the ratio $K = x/\bar{x}$, this area being shown shaded in Figure 5.9.

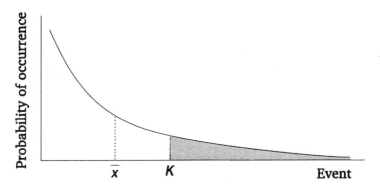

Figure 5.9 The exponential curve

As an example, suppose that the time between failures of a piece of equipment is found to vary exponentially. If results indicate that the mean time between failures is 1000 weeks, then what is the probability that the equipment will work for 700 weeks or more without a failure? Calculating K as $700/1000 = 0.7$ then from Table 5.2 the area beyond 0.7 is 0.497 which is the probability that the equipment will still be working after 700 weeks.

5.9.5 Weibull distribution

This is a continuous probability distribution and is given by Equation 5.43, where α is called the scale factor, β the shape factor and γ the location factor.

$$y = \alpha \beta (x - \gamma)^{\beta - 1} \exp\left(-\alpha (x - \gamma)^{\beta}\right) \quad (5.43)$$

The shape of the Weibull curve varies depending on the value of its factors. β is the most important, as shown in Figure 5.10, and the Weibull curve varies from an exponential ($\beta = 1.0$) to a normal distribution. In practice β varies from about $\frac{1}{3}$ to 5. Because the

Statistical analysis 69

Table 5.2 Area under the exponential curve from K to $+\infty$

K	0.00	0.02	0.04	0.06	0.08
0.0	1.000	0.980	0.961	0.942	0.923
0.1	0.905	0.887	0.869	0.852	0.835
0.2	0.819	0.803	0.787	0.771	0.756
0.3	0.741	0.726	0.712	0.698	0.684
0.4	0.670	0.657	0.644	0.631	0.619
0.5	0.607	0.595	0.583	0.571	0.560
0.6	0.549	0.538	0.527	0.517	0.507
0.7	0.497	0.487	0.477	0.468	0.458
0.8	0.449	0.440	0.432	0.423	0.415
0.9	0.407	0.399	0.391	0.383	0.375

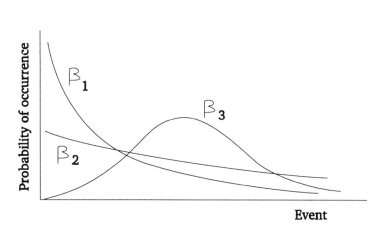

Figure 5.10 Weibull curves ($\alpha = 1$)

Weibull distribution can be made to fit a variety of different sets of data, it is popularly used for probability distributions.

Analytical calculations using the Weibull distribution are cumbersome. Usually predictions are made using Weibull probability paper. The data are plotted on this paper and the probability predictions read from the graph.

5.10 Sampling

A sample consists of a relatively small number of items drawn from a much larger population. This sample is analysed for certain attributes and it is then assumed that these attributes apply to the total population, within a certain tolerance of error.

Sampling is usually associated with the normal probability distribution and, based on this distribution, the errors which arise due to sampling can be estimated. Suppose a sample of n_s items is taken from a population of n_p items which are distributed normally. If the sample is found to have a mean of μ_s with a standard deviation of σ_s, then the mean μ_p of the population can be estimated to be within a certain tolerance of μ_s. It is given by Equation 5.44.

$$\mu_p = \mu_s \pm \frac{\gamma \sigma_s}{\sqrt{n_s}} \tag{5.44}$$

γ is found from the normal curve depending on the level of confidence we need in specifying μ_p. For $\gamma = 1$ this level is 68.26%, for $\gamma = 2$ it is 95.46% and for $\gamma = 3$ it is 99.73%.

Standard error of the mean σ_c is often defined as in Equation 5.45.

$$\sigma_c = \frac{\sigma_s}{\sqrt{n_s}} \tag{5.45}$$

Therefore Equation 5.44 can be re-written as in Equation 5.46.

$$\mu_p = \mu_s \pm \gamma \sigma_c \tag{5.46}$$

Statistical analysis 71

As an example, suppose that a sample of 100 items, selected at random from a much larger population, gives their mean weight as 20kg with a standard deviation of 100g. The standard error of the mean is therefore $100/(100)^{1/2} = 10$g and one can say with 99.73% confidence that the mean value of the population lies between $20 \pm 3 \times 0.01$ or 20.03kg and 19.97kg.

If in a sample of n_s items the probability of occurrence of a particular attribute is p_s, then the standard error of probability p_c is defined as in Equation 5.47, where $q_s = 1 - p_s$.

$$p_c = \left(\frac{p_s q_s}{n_s} \right)^{1/2} \tag{5.47}$$

The probability of occurrence of the attribute in the population is then given by Equation 5.48, where γ is again chosen to cover a certain confidence level.

$$p_p = p_s \pm \gamma p_c \tag{5.48}$$

As an example, suppose a sample of 500 items shows that 50 are defective. Then the probability of occurrence of the defect in the sample is $50/500 = 0.1$. The standard error of the probability is $(0.1 \times 0.9/500)^{1/2}$ or 0.0134. Therefore one can state with 95.46% confidence that the population from which the sample was drawn has a defect probability of $0.1 \pm 2 \times 0.0134$, i.e. 0.0732 to 0.1268; or one can state with 99.73% confidence that this value will lie between $0.1 \pm 3 \times 0.0134$, i.e. 0.0598 to 0.1402.

If two samples have been taken from the same population and these give standard deviations of σ_{s1} and σ_{s2} for sample sizes of n_{s1} and n_{s2} then Equation 5.45 can be modified to give the standard error of the difference between means as in Equation 5.49.

$$\sigma_{dc} = \left(\frac{\sigma_{s1}^2}{n_{s1}} + \frac{\sigma_{s2}^2}{n_{s2}} \right)^{1/2} \tag{5.49}$$

Similarly Equation 5.47 can be modified to give the standard error of the difference between probabilities of two samples from the same population as in Equation 5.50.

$$p_{dc} = \left(\frac{p_{s1} q_{s1}}{n_{s1}} + \frac{p_{s2} q_{s2}}{n_{s2}} \right)^{1/2} \tag{5.50}$$

5.11 Tests of significance

In taking samples one often obtains results which deviate from the expected. Tests of significance are then used to determine if this deviation is real or if it could have arisen due to sampling error.

5.11.1 Hypothesis testing

A hypothesis is set up and then tested at a given confidence level. For example, suppose a coin is tossed 100 times and it comes up heads 60 times. Is the coin biased or is it likely that this falls within a reasonable sampling error? The hypothesis is set up that the coin is not biased. Therefore one would expect that the probability of heads is 0.5, i.e. $p_s = 0.5$. The probability of tails, q_s is also 0.5. Using Equation 5.47 the standard error of probability is given by Equation 5.51.

$$p_c = \left(\frac{0.5 \times 0.5}{100} \right)^{1/2} = 0.05 \tag{5.51}$$

Therefore from Equation 5.48 the population probability at the 95.45% confidence level of getting heads is $0.5 + 2 \times 0.05 = 0.6$. Therefore it is highly likely that the coin is not biased and the results are due to sampling error.

The results of any significance test are not conclusive. For example, is 95.45% too high a confidence level to require? The higher the confidence level the greater the risk of rejecting a true hypothesis, and the lower the level the greater the risk of accepting a false hypothesis.

Suppose now that a sample of 100 items of production shows that five are defective. A second sample of 100 items is taken from the same production a few months later and gives two defectives. Does this show that the production quality is improving? Using Equation 5.50 the standard error of the difference between probabilities is given by expression 5.52.

$$\left(\frac{0.05 \times 0.95}{100} + \frac{0.02 \times 0.98}{100} \right)^{1/2} = 0.0259 \tag{5.52}$$

This is less than twice the difference between the two probabilities, i.e. $0.05 - 0.02 = 0.03$, therefore the difference is very likely to have arisen due to sampling error and it does not necessarily indicate an improvement in quality.

5.11.2 Chi-square test

This is written as χ^2. If O is an observed result and E is the expected result then Equation 5.53 is obtained.

$$\chi^2 = \sum \frac{(O-E)^2}{E} \tag{5.53}$$

The χ^2 distribution is given by tables such as Table 5.3, from which the probability can be determined. The number of degrees of freedom is the number of classes whose frequency can be assigned independently. If the data are presented in the form of a table having V vertical columns and H horizontal rows then the degrees of freedom are usually found as in Equation 5.54.

$$\text{Degrees of freedom} = (V-1)(H-1) \tag{5.54}$$

Returning to the earlier example, suppose a coin is tossed 100 times and it comes up heads 60 times and tails 40 times. Is the coin

Tests of significance

Table 5.3 The Chi-square distribution

Degrees of freedom	Probability level				
	0.100	0.050	0.025	0.010	0.005
1	2.71	3.84	5.02	6.63	7.88
2	4.61	5.99	7.38	9.21	10.60
3	6.25	7.81	9.35	11.34	12.84
4	7.78	9.49	11.14	13.28	14.86
5	9.24	11.07	12.83	15.09	16.75
6	10.64	12.59	14.45	16.81	18.55
7	12.02	14.07	16.01	18.48	20.28
8	13.36	15.51	17.53	20.09	21.96
9	14.68	16.92	19.02	21.67	23.59
10	15.99	18.31	20.48	23.21	25.19
12	18.55	21.03	23.34	26.22	28.30
14	21.06	23.68	26.12	29.14	31.32
16	23.54	26.30	28.85	32.00	34.27
18	25.99	28.87	31.53	34.81	37.16
20	28.41	31.41	34.17	37.57	40.00
30	40.26	43.77	46.98	50.89	53.67
40	51.81	55.76	59.34	63.69	66.77

biased? The expected values for heads and tails are 50 each so that expression 5.55 is obtained.

$$\chi^2 = \frac{(60-50)^2}{50} + \frac{(40-50)^2}{50} = 4 \qquad (5.55)$$

Table 5.4 The Chi-square distribution

Time (24 hour clock)	Number of accidents
0–6	9
6–12	3
12–18	2
18–24	6

The number of degrees of freedom is one since once we have fixed the frequency for heads that for tails is defined. Therefore entering Table 5.3 with one degree of freedom the probability level for $\chi^2 = 4$ is seen to be above 2.5% i.e. there is a strong probability that the difference in the two results arose by chance and the coin is not biased.

As a further example, suppose that over a 24 hour period the average number of accidents which occur in a factory is seen to be as in Table 5.4. Does this indicate that most of the accidents occur during the late night and early morning periods? Applying the χ^2 tests the expected value, if there was no difference between the time periods, would be the mean of the number of accidents, i.e. 5.

Therefore from Equation 5.53, expression 5.56 is obtained.

$$\chi^2 = \frac{(9-5)^2}{5} + \frac{(3-5)^2}{5} + \frac{(2-5)^2}{5} + \frac{(6-5)^2}{5} = 6 \qquad (5.56)$$

There are three degrees of freedom, therefore from Table 5.3 the probability of occurrence of the result shown in Table 5.4 is seen to be greater than 10%. The conclusion would be that although there is a trend, as yet there are not enough data to show if this trend is significant or not. For example, if the number of accidents were each three times as large, i.e. 27, 9, 6, 18 respectively, then χ^2 would be calculated as 20.67 and from Table 5.3 it is seen that this result is highly significant since there is a very low probability, less than 0.5%, that it can arise by chance.

5.11.3 Significance of correlation

The significance of the product moment correlation coefficient of Equation 5.19 or 5.20 can be tested at any confidence level by means of the standard error of estimation given by Equation 5.21. An alternative method is to use the Student t test of significance. This is given by Equation 5.57, where r is the correlation coefficient and n the number of items.

$$t = \frac{r(n-2)^{1/2}}{(1-r^2)^{1/2}} \tag{5.57}$$

Tables are then used, similar to Table 5.3, which give the probability level for $(n-2)$ degrees of freedom.

The Student t for the rank correlation coefficient is given by Equation 5.58, and the same Student t tables are used to check the significance of R.

$$t = R\left(\frac{n-2}{1-R^2}\right)^{1/2} \tag{5.58}$$

5.12 Bibliography

Besterfield, D.H. (1979) *Quality Control*, Prentice Hall.
Caplen, R.H. (1982) *A Practical Approach to Quality Control*, Business Books.
Chalk, G.O. and Stick, A.W. (1975) *Statistics for the Engineer*, Butterworths.
Cohen, S.S. (1988) *Practical Statistics*, Edward Arnold.

David, H.A. (1981) *Order Statistics*, Wiley.
Dudewicz, E.J. and Mishra, S.N. (1988) *Modern Mathematical Statistics*, Wiley.
Dunn, R.A. and Ramsing, K.D. (1981) *Management Science, a Practical Approach to Decision Making*, Macmillan.
Fitzsimmons, J.A. (1982) *Service Operations Management*, McGraw-Hill.
Grant, E.I. and Leavenworth, R.S. (1980) *Statistical Quality Control*, McGraw-Hill.
Hahn, W.C. (1979) *Modern Statistical Methods*, Butterworths.
Jones, M.E.M. (1988) *Statistics*, Schofield & Sims.
Mazda, F.F. (1979) *Quantitative Techniques in Business*, Gee & Co.
Siegel, A.F. (1988) *Statistics and Data Analysis*, Wiley.
Taylor, W.J. and Watling T.F. (1985) *The Basic Arts of Management*, Business Books.

6. Fourier analysis

6.1 Introduction

This chapter considers Fourier analysis and associated transform methods for both discrete-time and continuous-time signals and systems. Fourier methods are based on using real or complex sinusoids as basis functions, and they allow signals to be represented in terms of sums of sinusoidal components.

Euler and d'Alembert showed that the wave equation (Equation 6.1) could be satisfied by any well-behaved function of the variable $x + ct$ or $x - ct$, i.e. Equation 6.2.

$$\frac{\partial^2 \Phi}{\partial x^2} = \frac{1}{c^2} \frac{\partial^2 \Phi}{\partial t^2} \tag{6.1}$$

$$\Phi(x,t) = f(x + ct) + g(x - ct) \tag{6.2}$$

Boundary conditions, of course, impose restrictions on the functions f and g, but apart from these restrictions the functions f and g are completely arbitrary. This result started a long controversy, since it had been shown by Daniel Bernoulli (1727), that the solution to the wave equation could also be represented as a superposition of sinusoidal components. If the solution obtained by Bernoulli was as general as that of Euler and d'Alembert, it follows that an arbitrary function can be represented as a superposition of sinusoidal components. Euler had difficulty accepting this, and it was Jean-Baptiste Fourier (1768–1830) whose work 'Memoire sur la Chaleur' expounded the method now known as Fourier analysis and eventually showed the above equivalence.

It is of fundamental importance in both linear systems theory, and signal processing in general, that the input and/or output signals of

Fourier analysis

interest can be represented as linear combinations of simple basis functions, and the Fourier series representation is one such method that allows the signals of interest to be decomposed into harmonic components using a sinusoidal basis. This is clearly of interest if one is considering the spectral composition of signals, and in this area, Fourier analysis and the Fourier transform are used widely.

It should be realised that the Fourier representation is just one of many possible representations, and the choice of the signal representation will be determined by the particular nature of the problem under investigation.

Suppose that we have a function $f(t)$ that we wish to represent on a finite interval (t_1, t_2), where t can obviously represent time, space or any other dimension relevant to our problem, in terms of a set of a basis functions $\psi_1(t), \psi_2(t), ..., \psi_n(t)$. We will assume that these functions are orthogonal on the region of support (t_1, t_2), which is written as Equation 6.3.

$$\int_{t_1}^{t_2} \psi_i(t) \psi_j(t) \, dt = 0 \tag{6.3}$$

We will use the notation $<\psi_i|\psi_j>$ to denote the integral above. The idea of orthogonality expressed above, is the same as that applied to vectors and vector spaces, and our representation of $f(t)$ in terms of the functions $\psi_i(t)$, $i = 1, 2, ..., n$ is equivalent to representing a vector f in terms of an orthogonal set of vectors that span the space containing f.

We will assume that the representation of $f(t)$ can be written as a linear combination of the basis functions $\psi_i(t)$ as in Equation 6.4.

$$f(t) = \sum_{i=1}^{n} c_i \psi_i(t) \tag{6.4}$$

This will, in general, be in error since it is only a representation of the function $f(t)$, and we will require that the representation should be as close as possible to $f(t)$. Many measures of closeness exist, but a

80 Introduction

common measure is the mean squared error (MSE). We therefore require that the coefficients c_i, $i = 1, 2, ..., n$ are chosen to minimise its value in Equation 6.5.

$$MSE = \frac{1}{t_2 - t_1} \int_{t_1}^{t_2} \left[f(t) - \sum_{i=1}^{n} c_i \psi_i(t) \right]^2 dt \tag{6.5}$$

This is the mean squared error averaged over the interval (t_1, t_2). Defining the functions in Equations 6.6 and 6.7, then the expression for the mean square error above can be written as in Equation 6.8.

$$\alpha_i = \int_{t_1}^{t_2} f(t) \psi_i(t) \, dt = <f|\psi_i> \tag{6.6}$$

$$\beta_i = \int_{t_1}^{t_2} \psi_i^2 \, dt = <\psi_i|\psi_i> \tag{6.7}$$

$$MSE = \frac{1}{t_2 - t_1} \left[\int_{t_1}^{t_2} f(t)^2 \, dt + c_1^2 \beta_1 + c_2^2 \beta_1 + ... \right.$$
$$\left. + c_n^2 \beta_n - 2c_1 \alpha_1 - 2c_2 \alpha_2 - ... - 2c_n \alpha_n \right] \tag{6.8}$$

Using the identity of Equation 6.9, the expression for the mean squared error can be written as in Equation 6.10.

$$c_i^2 \beta_i - 2c_i \alpha_i = \left(c_i \sqrt{\beta_i} - \frac{\alpha_i}{\sqrt{\beta_i}} \right)^2 - \frac{\alpha_i^2}{\beta_i} \tag{6.9}$$

$$MSE = \frac{1}{t_2 - t_1} \left[\int_{t_1}^{t_2} f(t)^2 \, dt \right.$$
$$\left. + \sum_{i=1}^{n} \left(c_i \sqrt{\beta_i} - \frac{\alpha_i}{\sqrt{\beta_i}} \right)^2 - \sum_{i=1}^{n} \frac{\alpha_i^2}{\beta_i} \right] \tag{6.10}$$

From the form of this expression, it is clear that the MSE is always greater than or equal to zero, since it is the sum of squared terms, and that it achieves its least value when the middle term above is zero, i.e. as in Equation 6.11.

$$c_i = \frac{\alpha_i}{\beta_i} = \frac{<f|\psi_i>}{<\psi_i|\psi_i>} \tag{6.11}$$

Therefore, the best approximation of an arbitrary signal or function $f(t)$, in the mean squared error sense, over the region of support (t_1, t_2), that can be represented as a linear superposition of orthogonal basis functions, is achieved by choosing the coefficients c_i according to the above expression, which is the normalised projection of the function in the direction of the basis function ψ_i.

As is clear from above, when the coefficients c_i are chosen using the above criterion, the mean squared error is simply the averaged difference between two terms, and this difference can be made as small as one likes by including more and more terms in the summation term. Thus in the limit, the mean squared error is zero when Equation 6.12 is satisfied.

$$\int_{t_1}^{t_2} f(t)^2 \, dt = \sum_{i=1}^{\infty} c_i^2 \int_{t_1}^{t_2} \psi_i^2 \, dt \tag{6.12}$$

If this relationship holds, then the infinite sum of the weighted basis functions is said to converge in the mean to $f(t)$, an equality known as Parseval's relation, and if it holds for all $f(t)$ of a certain class, then the set $\psi_i(t)$ is said to be complete for that class of functions.

If the basis functions are complex valued functions of a real argument t, orthogonality is defined as in Equation 6.13 where ψ_j^* is the complex conjugate of ψ_j and the generalised Fourier coefficients are given by Equation 6.14.

82 Introduction

$$<\psi_i | \psi_j^*> = \int_{t_1}^{t_2} \psi_i \psi_j^* \, dt = 1 \; or \; 0 \tag{6.13}$$

$$c_i = \frac{<f | \psi_i>}{<\psi_i | \psi_i>} \tag{6.14}$$

If $<\psi_i | \psi_i> = 1$, the set ψ_i is called an orthonormal set of basis functions.

For example, the trigonometric system of functions with period 2π forms an orthogonal set over the region $(-\pi, \pi)$ and it can be shown that the following forms an orthonormal set over this region:

$$\frac{1}{\sqrt{2\pi}}, \frac{1}{\sqrt{\pi}} \sin(t), \frac{1}{\sqrt{\pi}} \cos(t), ..., \frac{1}{\sqrt{\pi}} \sin(nt), \frac{1}{\sqrt{\pi}} \cos(nt), ...$$

Clearly, other orthonormal sets or trigonometric functions can be defined over other intervals, e.g. for the interval $(0, \pi)$, the set

$$\left[\sqrt{\frac{2}{\pi}} \sin(nt) \right]_{n=1}^{\infty}$$

is orthonormal.

If ψ_i is an orthonormal set, the generalised Fourier coefficients for the representation of a function f, are given by Equation 6.15. As an example, we can consider the orthonormal trigonometric system of period 2π for which Equation 6.16 to 6.20 are satisfied, for $k = 2, 3, ...$

$$c_n = <f | \psi_n> \tag{6.15}$$

$$c_o = \frac{1}{\sqrt{2\pi}} \int_{-\pi}^{\pi} f(t) \, dt \tag{6.16}$$

$$c_1 = \frac{1}{\sqrt{\pi}} \int_{-\pi}^{\pi} f(t) \sin(t) \, dt \tag{6.17}$$

$$c_2 = \frac{1}{\sqrt{\pi}} \int_{-\pi}^{\pi} f(t) \cos(t) \, dt \tag{6.18}$$

$$c_{2k-1} = \frac{1}{\sqrt{\pi}} \int_{-\pi}^{\pi} f(t) \sin(kt) \, dt \tag{6.19}$$

$$c_{2k} = \frac{1}{\sqrt{\pi}} \int_{-\pi}^{\pi} f(t) \cos(kt) \, dt \tag{6.20}$$

Substituting these coefficients into the general expression for the expansion of $f(t)$ in terms of these orthonormal basis functions, we find that the Fourier series for the signal, or function, $f(t)$ can be expressed as in Equations 6.21 and 6.22, which may be written as in Equation 6.23, where A_0, A_n and B_n are appropriately defined.

$$\begin{aligned} f(t) = & \left[\frac{1}{\sqrt{2\pi}} \int_{-\pi}^{\pi} f(t) \, dt \right] \frac{1}{\sqrt{2\pi}} \\ & + \sum_{n=1}^{\infty} \left(\left[\frac{1}{\sqrt{\pi}} \int_{-\pi}^{\pi} f(t) \sin(nt) \, dt \right] \frac{1}{\sqrt{\pi}} \sin(nt) \right. \\ & \left. + \left[\frac{1}{\sqrt{\pi}} \int_{-\pi}^{\pi} f(t) \cos(nt) \, dt \right] \frac{1}{\sqrt{\pi}} \cos(nt) \right) \end{aligned} \tag{6.21}$$

$$\begin{aligned} f(t) = & \frac{1}{2} \left[\frac{1}{\pi} \int_{-\pi}^{\pi} f(t) \, dt \right] \\ & + \sum_{n=1}^{\infty} \left(\left[\frac{1}{\pi} \int_{-\pi}^{\pi} f(t) \sin(nt) \, dt \right] \sin(nt) \right. \\ & \left. + \left[\frac{1}{\pi} \int_{-\pi}^{\pi} f(t) \cos(nt) \, dt \right] \cos(nt) \right) \end{aligned} \tag{6.22}$$

$$f(t) = \frac{1}{2} A_0 + \sum_{n=1}^{\infty} (A_n \cos(nt) + B_n \sin(nt)) \tag{6.23}$$

What we have shown is that if a function $f(t)$ can be represented by a Fourier expansion, then the coefficients of the expansion may be calculated by the methods described. However, the question of convergence of the expansion has not yet been addressed. The necessary and sufficient conditions for the convergence of a Fourier expansion are well known, but it is useful to state a sufficient condition, known as the Dirichlet condition, which states that if $f(t)$ is bounded and of period T and if $f(t)$ has at most a finite number of maxima and minima in one period and a finite number of discontinuities, then the Fourier series for $f(t)$ converges to $f(t)$ at all points where $f(t)$ is continuous, and converges to the average of the right hand and left hand limits of $f(t)$ at each point where $f(t)$ is discontinuous.

In many applications, such as the design of filters, it may be necessary to use only a finite number of terms of the Fourier series to approximate a function $f(t)$ over $(0, T_p)$, and it is therefore of much interest to enquire what effect the truncation of the series has (Banks, 1990). The error incurred obviously decreases as one takes into account more and more terms when the function $f(t)$ is continuous. However, in the neighbourhood of discontinuities, ripples occur, the magnitude of which remains roughly the same even as more and more terms are included in the Fourier expansion. These ripples are referred to as Gibbs' oscillations.

6.2 Generalised Fourier expansion

As an example of using the generalised Fourier expansion, let us consider representing the function shown in Figure 6.1 in terms of the Legendre polynominals, which form an orthogonal set on the interval $(-1,1)$. The corresponding normalised basis functions are given by Equations 6.24 to 6.27, where the $P_{n(t)}$ are the Legendre polynominals, which may be generated by the expression of Equation 6.28, or by using the recurrence relation of Equation 6.29.

$$\Phi_0(t) = \frac{1}{\sqrt{2}} \tag{6.24}$$

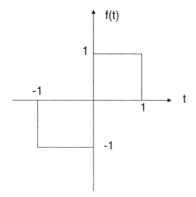

Figure 6.1 The function f(t)

$$\Phi_1(t) = t\sqrt{\frac{3}{2}} \tag{6.25}$$

$$\Phi_2(t) = \sqrt{\frac{5}{2}}\left(\frac{3}{2}t^2 - \frac{1}{2}\right) \tag{6.26}$$

$$\Phi_n(t) = \left(\frac{2n+1}{2}\right)^{1/2} P_n(t) \tag{6.27}$$

$$P_n(t) = \frac{1}{2^n n!}\frac{d^n}{dt^n}(t^2-1)^n \tag{6.28}$$

$$nP_n(t) = (2n-1)tP_{n-1}(t) - (n-1)P_{n-2}(t) \tag{6.29}$$

The generalised Fourier coefficient is given by Equation 6.30, where $n = 0, 1, 2...$

$$c_n = <f|\Phi_n> = \int_{-1}^{1} f(t)\Phi_n(t)\,dt \tag{6.30}$$

Using the expressions for Φ_n given above, we obtain Equations 6.31 to 6.34.

$$c_0 = \int_{-1}^{1} \frac{f(t)}{\sqrt{2}} \, dt = 0 \qquad (6.31)$$

$$c_1 = \int_{-1}^{1} \sqrt{\frac{3}{2}} \, t f(t) \, dt = \sqrt{\frac{3}{2}} \qquad (6.32)$$

$$c_2 = \int_{-1}^{1} \sqrt{\frac{5}{2}} \left(\frac{3}{2} t^2 - \frac{1}{2} \right) f(t) \, dt = 0 \qquad (6.33)$$

$$c_3 = \int_{-1}^{1} \sqrt{\frac{7}{2}} \left(\frac{5}{2} t^3 - \frac{3}{2} t \right) f(t) \, dt = \sqrt{\frac{7}{16}} \qquad (6.34)$$

In general, the coefficient c_n is zero when n is even, for this particular example. Hence the function $f(t)$ in Figure 6.1 can be represented in terms of the Legendre polynominals as in Equation 6.35.

$$\begin{aligned} f(t) &= \sum_{k=0}^{\infty} c_k \Phi_k(t) \\ &= \frac{3}{2} t + \frac{7}{4\sqrt{2}} \left(\frac{5}{2} t^3 - \frac{3}{2} t \right) + \ldots + c_n \Phi_n(t) + \ldots \end{aligned} \qquad (6.35)$$

We have shown that an arbitrary function, $f(t)$, can be expressed in terms of a superposition of orthogonal basis functions, and we have derived expressions for the generalised Fourier coefficients in terms of the projection of the given function onto the orthogonal basis functions.

If a signal, $x(t)$, repeats itself exactly every T_p seconds, then $x(t)$ may be represented as a linear combination of harmonically related-complex exponentials of the form given in Equation 6.36, where the fundamental frequency is given by Equation 6.37.

Fourier analysis

$$x(t) = \sum_{k=-\infty}^{\infty} c_k e^{j 2 \pi k f_0 t} \tag{6.36}$$

$$f_0 = \frac{1}{T_p} \tag{6.37}$$

Hence one can regard the exponential signals:

$$e^{j 2 \pi k f_0 t} \qquad k = 0, 1, 2, \ldots$$

as building blocks for periodic signals of various forms constructed by choosing the fundamental frequency and the coefficients c_k.

A periodic signal $x(t)$ with period T_p may represented by a Fourier series where f_0 is selected to be the reciprocal of T_p. The Fourier coefficients c_k are obtained by multiplication of the Fourier representation by the complex exponential $e^{-j 2 \pi l f_0 t}$ where l is an integer, followed by integration over a single period, either 0 to T_p or more generally t_0 to $t_0 + T_p$ where t_0 is arbitrary (Equation 6.38).

$$\begin{aligned}
& \int_{t_0}^{t_0 + T_p} x(t) e^{-j 2 \pi l f_0 t} dt \\
&= \int_{t_0}^{t_0 + T_p} e^{-j 2 \pi l f_0 t} \left(\sum_{k=-\infty}^{\infty} c_k e^{j 2 \pi k f_0 t} \right) dt \\
&= \sum_{k=-\infty}^{\infty} c_k \int_{t_o}^{t_o + T_p} e^{j 2 \pi f_o (k-l) t} dt
\end{aligned} \tag{6.38}$$

The integral on the r.h.s. is identically zero if $k \neq l$ and is equal to T_p if $k = l$, and therefore Equation 6.39 can be obtained.

$$c_k = \frac{1}{T_p} \int_{t_0}^{t_0 + T_p} x(t) e^{-j 2 p i k f_0 t} dt \tag{6.39}$$

It is now of interest to consider the average power, P, of a periodic signal $x(t)$ which is given by Equation 6.40.

$$P = \frac{1}{T_p} \int |x(t)|^2 \, dt = \frac{1}{T_p} \int x(t) x^*(t) \, dt \qquad (6.40)$$

Using the Fourier representation of $x(t)$ and the expression found for the Fourier coefficients, we can write Equation 6.41.

$$\begin{aligned} P &= \frac{1}{T_p} \int x(t) \left(\sum_{k=-\infty}^{\infty} c_k^* e^{-2j\pi k f_0 t} \right) dt \\ &= \sum_{k=-\infty}^{\infty} c_k^* \left(\frac{1}{T_p} \int x(t) e^{-2j\pi k f_0 t} \right) dt \\ &= \sum_{k=-\infty}^{\infty} c_k^* c_k = \sum_{k=-\infty}^{\infty} |c_k|^2 \end{aligned} \qquad (6.41)$$

This relationship between the average power and the square of the Fourier coefficients is called Parseval's relation, as mentioned earlier in the context of function representation. The kth harmonic component of the signal has a power $|c_k|^2$ and hence the total average power in a periodic signal is just the sum of the average powers in all the harmonics. If $|c_k|^2$ is plotted as a function of kf_0, the resulting function is called the power density spectrum of the periodic signal $x(t)$. Since the power in a periodic signal exists only at discrete values of frequencies, i.e. Equation 6.42 is satisfied, the signal is said to have a line spectrum, and the frequency spacing between two adjacent spectral lines is equal to the reciprocal of the fundamental period T_p.

$$f = 0, \ \pm f_0, \ \pm 2f_0, \ \dots \qquad (6.42)$$

The Fourier coefficients c_{subk} are in general complex, and may be represented as in Equation 6.43 and hence an alternative to the power spectrum representation may be obtained by plotting the magnitude spectrum $|c_k|$ and phase spectrum Θ_k as a function of frequency.

$$c_k = |c_k| e^{j\Theta_k} \tag{6.43}$$

Phase information is therefore lost in using the power spectral density.

For a real valued periodic signal, the Fourier coefficients c_k can easily be shown to satisfy Equation 6.44, which implies that the power spectrum, and hence the magnitude spectrum, are symmetric functions of frequency. However, the phase spectrum is an odd function.

$$c_{-k} = c_k^* \tag{6.44}$$

As an example of this analysis, the Fourier series and power spectral density of the rectangular pulse train shown in Figure 6.2 will be given, and since teh signal $x(t)$ is even, we will select the integration interval from $\frac{-T^p}{2}$ to $\frac{T_p}{2}$, as in Equation 6.45 and 6.46, for $k = \pm 1, = \pm 2, \ldots$, and the power spectral density is obtained by squaring these quantities.

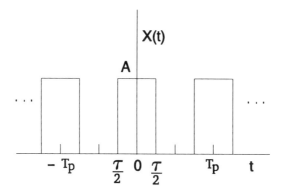

Figure 6.2 Continuous time periodic train of rectangular pulses

$$c_0 = \frac{1}{T_p} \int_{-\tau/2}^{\tau/2} x(t)\,dt = \frac{A\tau}{T_p} \qquad (6.45)$$

$$\begin{aligned}
c_k &= \frac{1}{T_p} \int_{-\tau/2}^{\tau/2} A\, e^{-j2\pi k f_0 t}\,dt \\
&= \frac{A}{\pi f_0 k T_p} \frac{e^{j\pi k f_0 \tau} - e^{-j\pi k f_0 \tau}}{2j} = \frac{A\tau}{T_p} \frac{\sin(\pi k f_0 \tau)}{\pi k f_0 \tau}
\end{aligned} \qquad (6.46)$$

It is now important to introduce the concept of aperiodic signals and transform methods, before dealing with discrete data.

6.3 Fourier transforms

Consider an aperiodic signal $x(t)$ which is of finite duration, Figure 6.3. A periodic signal $x_p(t)$ can be created from $x(t)$ by translation by

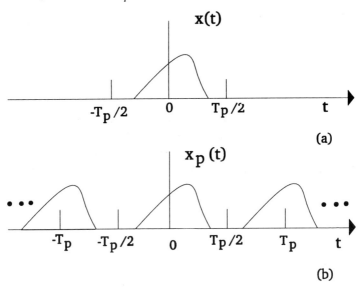

Figure 6.3 An aperiodic signal and a periodic signal constructed by repeating the periodic signal

fixed amounts, T_p, as shown. This new periodic signal, $x_p(t)$, approaches $x(t)$ in the limit as T_p approaches infinity, and the spectrum of $x_p(t)$ should be obtainable from the spectrum of $x(t)$ in the same limit.

As before, we can obtain Equations 6.47 and 6.48, where $f_0 = 1/T_p$.

$$x_p(t) = \sum_{k=-\infty}^{\infty} c_k e^{j2\pi k f_0 t} \tag{6.47}$$

$$c_k = \frac{1}{T_p} \int_{-T_p/2}^{T_p/2} x_p(t) e^{-j2\pi k f_0 t} dt \tag{6.48}$$

Between the limits $\pm \frac{T_p}{2}$, $x_p(t)$ can be replaced by $x(t)$, and since $x(t)$ is zero outside this range of integration, Equation 6.49 may be obtained.

$$c_k = \frac{1}{T_p} \int_{-\infty}^{\infty} x(t) e^{-j2\pi k f_0 t} dt \tag{6.49}$$

A new function, $X(f)$, can now be defined as the Fourier transform of $x(t)$ as in Equation 6.50 and the Fourier coefficient c_k can be written as in Equation 6.51, which are samples of $X(f)$ taken at multiples of f_0 and also scaled by the factor f_0 (= $1/T_p$).

$$X(f) = \int_{-\infty}^{\infty} x(t) e^{-j2\pi f t} dt \tag{6.50}$$

$$c_k = \frac{1}{T_p} X(k f_0) \tag{6.51}$$

Therefore Equation 6.52 can be obtained.

$$x_p(t) = \frac{1}{T_p} \sum_{k=-\infty}^{\infty} X\left(\frac{k}{T_p}\right) e^{j2\pi kt/T_p} \qquad (6.52)$$

As described above, we require to take the limit as T_p approaches infinity, and it is therefore convenient to define a frequency differential $\delta f = \frac{1}{T_p}$ such that in the limit δf approaches zero. We may therefore write Equation 6.53, and taking the limit as $\delta f \to \sim 0$ we obtain Equations 6.54 and 6.55.

$$x_p(t) = \sum_{k=-\infty}^{\infty} X(k\delta f) e^{j2\pi kt\delta f} \delta f \qquad (6.53)$$

$$\lim_{T_p \to \infty} x_p(t) = x(t) = \lim_{\delta_f \to 0} \sum_{k=-\infty}^{\infty} X(k\delta f) e^{j2\pi ft} df \qquad (6.54)$$

$$x(t) = \int_{-\infty}^{\infty} X(f) e^{j2\pi ft} df \qquad (6.55)$$

These equations show that $x(t)$ and $X(f)$ form a so-called Fourier Transform pair, and the Fourier Transform of $x(t)$ exists if the signal has a finite energy, i.e. as in Equation 6.56.

$$\int_{-\infty}^{\infty} |x(t)|^2 dt < \infty \qquad (6.56)$$

Earlier we defined the average power of a periodic signal. It is now possible similarly to define the energy of an aperiodic signal as in Equations 6.57 and 6.58.

$$E = \int_{-\infty}^{\infty} |x(t)|^2 dt = \int_{-\infty}^{\infty} x(t) x^*(t) dt$$
$$= \int_{-\infty}^{\infty} x(t) dt \left(\int_{-\infty}^{\infty} X^*(f) e^{-j2\pi ft} df \right) \qquad (6.57)$$

$$E = \int_{-\infty}^{\infty} X^*(f) \, df \left(\int_{-\infty}^{\infty} x(t) \, e^{-j2\pi ft} \, dt \right)$$
$$= \int_{-\infty}^{\infty} |X(f)|^2 \, df \tag{6.58}$$

Hence the energy of an aperiodic signal can be written as in Equation 6.59 which is also known as Parseval's relation (Wax, 1954).

$$E = \int_{-\infty}^{\infty} |x(t)|^2 \, dt = \int_{-\infty}^{\infty} |X(f)|^2 \, df \tag{6.59}$$

6.4 Discrete sequences

In order for a digital computer to manipulate a signal, the signal must be sampled at a chosen sampling rate, $1/T_s$, giving rise to a set of numbers called a sequence. If the continuous signal was $x(t)$, the sampled sequence is represented by $x(nT_s)$, where n is an integer, and the independent variable, t, could represent time or a spatial co-ordinate for example. The analysis can, of course, be extended into higher dimensions.

In order to move between the continuous domain and the discrete domain, the idea of a sampling function must be introduced, and from the definition of the Dirac delta function ($\delta(t)$) as in Equation 6.60, it is clear that this function meets the requirement.

$$\int_{-\infty}^{\infty} x(t) \, \delta(t - \tau) \, dt = x(\tau) \tag{6.60}$$

Consider an analog signal $x(t)$ that is continuous in both time and amplitude, and assume that $x(t)$ has infinite duration but finite energy. Let the sample values of the signal $x(t)$ at times $t = 0, \pm T_s, \pm 2T_s, \ldots$, be denoted by the series $x(nT_s)$, $n = 0, \pm 1, \pm 2, \ldots$, where T_s is the sampling period and $f_s = 1/T_s$, is the sampling rate. The discrete time signal $x_d(t)$ that is obtained by sampling the continuous signal $x(t)$ can then be written as in Equation 6.61, where $\delta(t - nT_s)$ is a Dirac delta function located at time $t = nT_s$ and each delta function is weighted by the corresponding sample value of the input signal $x(t)$.

$$x_d(t) = \sum_{n=-\infty}^{\infty} x(nT_s) \delta(t - nT_s) \tag{6.61}$$

For the case of discrete data, exactly analogous transform methods may be applied as for continuous signals, and we must now discuss the implementation issues for such methods.

6.5 The discrete Fourier transform

The discrete Fourier transform (DFT) is used extensively in digital signal processing (Burrus and Parks, 1985), and is used routinely for the detection and estimation of periodic signals. The DFT of a discrete-time signal $x(n)$ is defined as in Equation 6.62, where $k = 0, 1,..., N - 1$ and $W_N^{nk} = e^{j2\pi nk/N}$ are the basis functions of the DFT.

$$X(k) = \frac{1}{N} \sum_{n=0}^{N-1} x(n) W_N^{nk} \tag{6.62}$$

These functions are sometimes known as 'twiddle factors'. The basis functions are periodic and define points on the unit circle in the complex plane. Figure 6.4 illustrates the cyclic property of the basis functions for an eight point DFT, and the basis functions are equally spaced around the unit circle at frequency increments of F/N, where F is the sampling rate of the input signal sequence. In this figure the cyclic character of the twiddle factors are illustrated as follows:

$$W_8^0 = W_8^8 = W_8^{16} = W_8^{24} = \ldots$$
$$W_8^1 = W_8^9 = W_8^{17} = W_8^{25} = \ldots$$

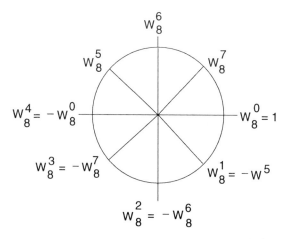

Figure 6.4 Cyclic properties of the basis functions for an eight point DFT

$$W_8^2 = W_8^{10} = W_8^{18} = W_8^{26} = \ldots$$

$$\vdots \qquad \vdots \qquad \vdots \qquad \vdots$$

$$W_8^7 = W_8^{15} = W_8^{23} = W_8^{31} = \ldots$$

The set of frequency samples which define the spectrum $X(k)$, are given on a frequency axis whose discrete frequency locations are given by Equation 6.63 where $k = 0, 1, \ldots, N - 1$.

$$f_k = k\frac{F}{N} \qquad (6.63)$$

The frequency resolution of the DFT is equal to the frequency increment F/N and is referred to as the bin spacing of the DFT outputs. The frequency response of any DFT bin output is determined by applying a complex exponential input signal and evaluating the DFT bin output response as the frequency is varied.

96 The discrete Fourier transform

Consider an input signal given by Equation 6.64, then the DFT of $x(n)$ can be expressed as a function of the arbitrary frequency variable f by Equation 6.65.

$$x(n) = e^{j2\pi fn/F} \tag{6.64}$$

$$X(k) = \frac{1}{N}\sum_{n=0}^{N-1} x(n)\, W_N^{nk} = \frac{1}{N}\sum_{n=0}^{N-1} e^{j2\pi fn/F}\, W_N^{nk} \tag{6.65}$$

This summation can be evaluated using the geometric series summation to give Equation 6.66.

$$X(k) = \frac{1}{N}\frac{1 - e^{-j2\pi(k/N - f/F)N}}{1 - e^{-j2\pi(k/N - f/F)}} \tag{6.66}$$

Defining Ω by Equation 6.67, the DFT of $x(n)$ can be written as in Equation 6.68.

$$\Omega = 2\pi\left(\frac{k}{N} - \frac{f}{F}\right) \tag{6.67}$$

$$X(k) = e^{-j\Omega(N-1)/2}\,\frac{\sin\dfrac{\Omega N}{2}}{N\sin\dfrac{\Omega}{2}} \tag{6.68}$$

The first term in this expression is the phase of the response, and the ratio of sines is the amplitude response. If k is an integer, then $X(k) = 1$, for these values of k, and zero elsewhere. If k is not an integer, then none of the DFT values are zero. This is called spectral leakage. Hence, a unit impulse at the kth frequency location is only obtained when Equation 6.69 is satisfied, where k is an integer.

$$f_k = \frac{kF}{N} \tag{6.69}$$

Fourier analysis 97

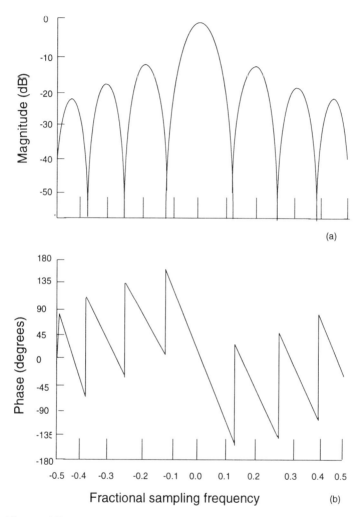

Figure 6.5 Frequency and phase response of a DFT bin output for an eight point FFT: (a) magnitude; (b) phase angle

Figure 6.5 shows the frequency response of a DFT bin output for an eight point DFT, and the high sidelobe levels are a consequence of truncation.

6.6 The inverse discrete Fourier transform

Given the discrete Fourier transform $X(k)$ of the input sequence $x(n)$, the inverse discrete Fourier transform (IDFT) of $X(k)$ is the original time sequence, and is given by Equation 6.70.

$$x(n) = \sum_{k=0}^{N-1} X(k) W_n^{-nk} \tag{6.70}$$

The proof is easily obtained by substitution and interchanging the order of summation. The form of the IDFT is the same as the DFT apart from the factor $\frac{1}{N}$ and the sign change in the exponent of the basis functions, and consequently the DFT algorithm can be used to compute these transforms in either direction. It should be noted that the DFT is equal to the Z-transform of a sequence, $x(n)$, evaluated at equally spaced inputs on the unit circle in the z-plane (see later).

6.7 The fast Fourier transform

Special analysis and many other applications often require DFTs to be performed on data sequences in real time and on contiguous sets of input samples. If the input sequence has N samples, the computation of the DFT requires N^2 complex multiplies and $N^2 - N$ complex additions.

The FFT is a fast algorithm for the efficient implementation of the DFT where the number of time samples of the input signal N are transformed into N frequency points, and the required number of arithmetic operations is reduced to the order of $\frac{N}{2} \log_2(N)$. Several approaches can be used to develop the FFT algorithm.

Fourier analysis

Starting with the DFT expression, consider factorising it into two DFTs of, length $N/2$ by splitting the input samples into even and odd samples, in Equations 6.71 and 6.72, where $k = 0, 1, 2,..., N - 1$.

$$X(k) = \frac{1}{N} \sum_{n=0}^{N-1} x(n) W_N^{nk} \tag{6.71}$$

$$X(k) = \frac{1}{N} \sum_{m=0}^{N/2-1} x(2m) W_N^{2mk} + \frac{1}{N} \sum_{m=0}^{N/2-1} x(2m+1) W_N^{(2m+1)k} \tag{6.72}$$

x_1 and x_2 represent the normalised even and odd sample components, as in Equations 6.73 and 6.74, for $m = 0, 1, 2, ... (N/2 - 1)$.

$$x_1(m) = \frac{x(2m)}{N} \tag{6.73}$$

$$x_2(m) = \frac{x(2m+1)}{N} \tag{6.74}$$

Therefore Equation 6.72 may be written as in Equation 6.75, where Equation 6.76 holds and each of the summation terms is reduced to an $N/2$ point DFT.

$$X(k) = \sum_{m=0}^{N/2-1} x_1(m) W_{N/2}^{mk} + W_N^k \sum_{m=0}^{N/2-1} x(m) W_{N/2}^{mk} \tag{6.75}$$

$$W_N^{2n} = W_{N/2}^n \tag{6.76}$$

The general form of the algorithm may be written as in Equation 6.77 and 6.78, where Equations 6.79 and 6.80 hold.

$$X(k) = X_1(k) + W_N^k X_2(k) \tag{6.77}$$

$$X\left(k + \frac{N}{2}\right) = X_1(k) + W_N^{k+N/2} X_2(k)$$

$$= X_1(k) - W_N^k X_2(k) \tag{6.78}$$

$$W_N^{k+N/2} = -W_N^k \tag{6.79}$$

$$W_{N/2}^{m(k+N/2)} = W_{N/2}^{mk} \tag{6.80}$$

Since the DFT output is periodic Equations 6.81 and 6.82 can be obtained and the form of the algorithm given above is refered to as the decimation in time FFT butterfly, and an example for a sixteen point FFT is shown in Figures 6.6 and 6.7

$$X_1(k) = X_1\left(k + \frac{N}{2}\right) \tag{6.81}$$

$$X_2(k) = X_2\left(k + \frac{N}{2}\right) \tag{6.82}$$

The decomposition process is repeated until two-point DFTs are generated. Each decomposition is called a stage, and the total number of stages is given by Equation 6.83.

$$M(N) = \log_2 N \tag{6.83}$$

Thus a 16 point DFT requires 4 stages, as shown in Figure 6.8. The algorithm is now re-applied to each of the $N/2$ sample DFTs.

Assuming that $N/2$ is even, the same process can be carried out on each of the $N/2$ point DFTs to further reduce the computation.

If $N = 2^M$ then the whole process can be repeated M times to reduce the computation to that of evaluating N single point DFTs.

Fourier analysis 101

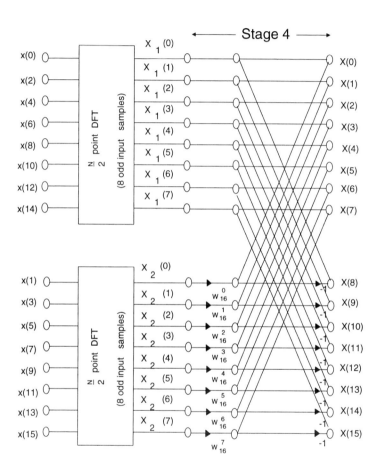

Figure 6.6 First step in developing a 16 point decimation in time FFT signal flow graph

102 The fast Fourier transform

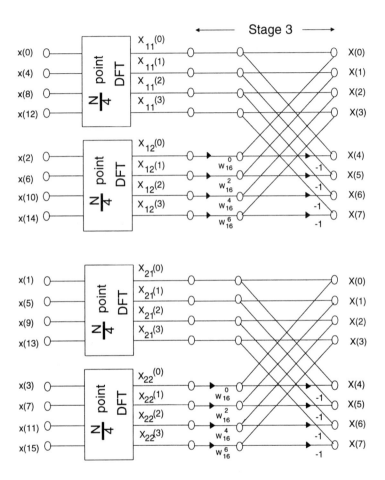

Figure 6.7 Second step in developing a 16 point decimation in time FFT signal flow graph

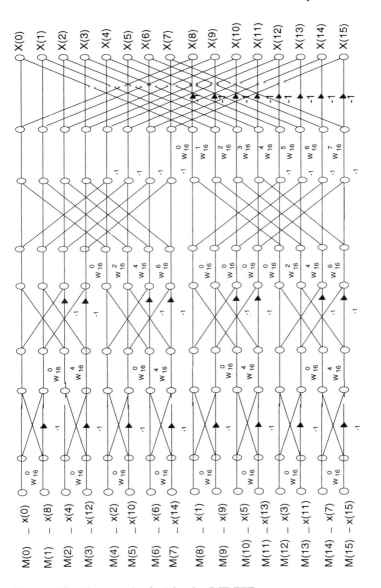

Figure 6.8 Flow graph of a 16 point DIT FFT

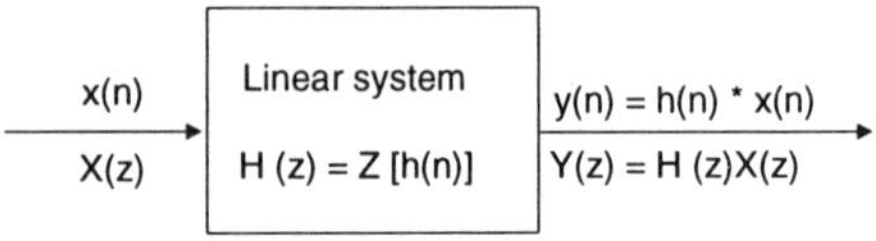

104 Linear time-invariant digital systems

Figure 6.9 Transfer function relationship for linear systems

6.8 Linear time-invariant digital systems

The theory of discrete-time, linear, time-invariant systems forms the basis for digital signal processing, and a discrete-time system performs an operation on the input signal according to a defined criteria to produce a modified output signal. The input signal $x(n)$, is the system excitation, and $y(n)$ is the response of the system to the excitation, as shown in Figure 6.9, (Kamen, 1990).

The transformation operation can be represented by the operator R. Linear and time-invariant systems can be completely characterised by their impulse response, $h(n)$, which is defined by Equation 6.84.

$$h(n) = R[\delta(n)] \tag{6.84}$$

Once the impulse response is determined, the output of the system for any given input signal is obtained by convolving the input with the impulse response of the system as in Equation 6.85. A system is linear if and only if the system's response to the sum of two signals, each multiplied by arbitrary scalar constants, is equal to the sum of the system's responses to the two signals taken separately, as in Equation 6.86.

$$y(n) = R[x(n)] = \sum_{k=-\infty}^{\infty} x(k) h(n-k) \tag{6.85}$$

$$\begin{aligned} y(n) &= R[a_1 x_1(n) + a_2 x_2(n)] \\ &= a_1 R[x_1(n)] + a_2 R[x_2(n)] \end{aligned} \tag{6.86}$$

Fourier analysis

A system is time-invariant if the response to a shifted version of the input is identical to a shifted version of the response based on the unshifted input, and a system operator R is time-invariant if Equation 6.87 holds for all values of m.

$$R[x(n-m)] = z^{-m} R[x(n)] \qquad (6.87)$$

The operator z^{-m} represents a signal delay of m samples.

The z-transform provides a very powerful method for the analysis of linear time-invariant discrete systems, which can be represented in terms of difference equations, via the operator R, relating the input signal samples to the output signal samples.

In the following, we will define the properties of the z-transform and then we will apply the methods to the analysis of discrete time linear time-invariant systems.

The z-transform of any data sequence $x(n)$, is defined as in Equation 6.88.

$$X(z) = \sum_{k=-\infty}^{\infty} x_k z^{-k} \qquad (6.88)$$

In this definition, z is a continuous complex variable and $X(z)$ is referred to as the 'two-sided z-transform', since both positive and negative values of the index k are allowed.

As an example of a z-transform, let us consider $x(n)$ to be samples of an exponential function such that $x_k = 0$ for $k < 0$, and $e^{-\alpha k}$ for $k \geq 0$ and $\alpha > 0$. The z-transform of this sampled signal is given by Equation 6.89, which is a simple rational function of α and z. Since the signal is zero for $k < 0$, we have a one sided transform in this case. Table 6.1 shows some common sampled sequences and their z-transforms.

$$X(z) = \sum_{k=0}^{\infty} e^{-\alpha k} z^{-k} = \frac{z}{z - e^{-\alpha}} \qquad (6.89)$$

106 Linear time-invariant digital systems

Table 6.1 Commonly used z-transform pairs

$x(n)$	$X(z)$	Region of convergence
$\delta(n)$	1	All z
$u(n)$	$\dfrac{z}{z-1}$	$|z|>1$
$a^n u(n)$	$\dfrac{z}{z-a}$	$|z|>|a|$
$-a^n u(-n-1)$	$\dfrac{z}{z-a}$	$|z|<|a|$
$n a^n u(n)$	$\dfrac{az}{(z-a)^2}$	$|z|>|a|$
$-n a^n u(-n)$	$\dfrac{az}{(z-a)^2}$	$|z|<|a|$
$\dfrac{(n+k-1)!}{N!(k-1)} a^n u(n)$	$\dfrac{z^k}{(z-a)^k}$	$|z|>|a|$
$a^{|n|}$	$\dfrac{a^2-1}{a} \dfrac{z}{(z-a)(z-1/a)}$	$|a|<|z|$ $<|a|^{-1}$
$\sin(n\omega) u(n)$	$\dfrac{z\sin\omega}{z^2-2z\cos\omega+1}$	$|z|>1$
$\cos(n\omega) u(n)$	$\dfrac{z(z-\cos\omega)}{z^2-2z\cos\omega+1}$	$|z|>1$
$a^n \sin(n\omega) u(n)$	$\dfrac{az\sin\omega}{z^2-2az\cos\omega+a^2}$	$|z|>|a|$
$a^n \cos(n\omega) u(n)$	$\dfrac{z(z-a\cos\omega)}{z^2-2az\cos\omega+a^2}$	$|z|>|a|$

In linear systems analysis, the concept of a transfer function is fundamental and is defined simply as the transform of the output of the system divided by the transform of the input. The Laplace transform is an important tool that is used in continuous systems theory, and for a continuous signal $x(t)$, this transform is defined as in Equation 6.90.

$$X(s) = \int_0^\infty x(t) e^{-st} dt \tag{6.90}$$

If, as before, we sample this continuous signal $x(t)$, we obtain the sampled signal $x_s(t)$ given by Equation 6.91, with $x(t) = 0$ for $t < 0$.

$$x_s(t) = \sum_{n=0}^\infty x(t) \delta(t - nT) \tag{6.91}$$

The Laplace transform of the sampled signal $x_s(t)$ is then as in Equation 6.96.

$$\begin{aligned} X_s(s) &= \int_0^\infty \sum_{n=0}^\infty x(t) \delta(t - nT) e^{-st} dt \\ &= \sum_{n=0}^\infty x(nT) e^{-snT} \end{aligned} \tag{6.92}$$

Making the substitution gives Equation 6.93.

$$z = e^{sT} \tag{6.93}$$

Using the definition of Equation 6.94 gives Equation 6.95 which is the z-transform of the sampled sequence, $x(nT)$ and is also the Laplace transform of the sampled signal $x_s(t)$.

$$X(z) = X_s(s) \tag{6.94}$$

$$X(z) = \sum_{n=0}^{\infty} x(nT) z^{-n} \qquad (6.95)$$

Before deriving an expression for the transfer function and frequency response of a general linear system, a few of the properties of the z-transform need to be explained.

6.8.1 Linearity

The z-transform of the sum of two sequences multiplied by arbitrary constants is the sum of the z-transforms of the individual sequences, as in Equation 6.96, where Z represents the z-transform operator and $X(z)$ and $Y(z)$ are the z-transforms of the sequences $x(n)$ and $y(n)$ respectively.

$$Z[ax(n) + by(n)] = aX(z) + bY(z) \qquad (6.96)$$

6.8.2 Delay property

Consider a sampled sequence that has been delayed by p samples, and define the delayed sequence to be $x'(nT)$, as in Equation 6.97.

$$x'(nT) = x(nT - pT) \qquad (6.97)$$

The z-transform of this delayed sequence is then as in Equation 6.98.

$$\begin{aligned} X'(z) &= \sum_{n=0}^{\infty} x'(n) z^{-n} \\ &= \sum_{n=0}^{p-1} x'(n) z^{-n} + \sum_{n=p}^{\infty} x'(n) z^{-n} \end{aligned} \qquad (6.98)$$

Fourier analysis

This may be written in terms of the undelayed sequence, and with a redefinition of the summation index in the second term, we obtain Equation 6.99.

$$X'(z) = \sum_{n=0}^{p-1} x(n-p) z^{-n} + \sum_{n=0}^{\infty} x(n) z^{-(n+p)} \qquad (6.99)$$

Therefore, the z-transform of a delayed sequence is given by Equation 6.100.

$$X'(z) = \sum_{n=0}^{p-1} x(n-p) z^{-n} + z^{-p} X(z) \qquad (6.100)$$

The first term represents a contribution from any initial conditions, and if $x(n) = 0$ for $n < 0$, the z-transform of a delayed sequence is just $z^{-p} X(z)$, where p is the delay. This is of course, analogous to the shift theorem for the Laplace transform.

6.8.3 Convolution summation property

From the linear systems theory point of view, this represents one of the most valuable properties. Consider the definition of the convolution of an input sequence $x(n)$, with the impulse response $h(k)$ of a linear system, to yield an output $y(n)$, as in Equation 6.101.

$$y(n) = \sum_{k=-\infty}^{\infty} h(k) x(n-k) \qquad (6.101)$$

Taking the double sided z-transform of this equation gives Equation 6.106.

$$Y(z) = \sum_{n=-\infty}^{\infty} \left[\sum_{k=-\infty}^{\infty} h(k) x(n-k) \right] z^{-n} \qquad (6.102)$$

Again, changing the order of summation, and defining $p = n - k$, we obtain Equation 6.103.

$$Y(z) = \sum_{k=-\infty}^{\infty} h(k) z^{-k} \sum_{p=-\infty}^{\infty} x(p) z^{-p} \qquad (6.103)$$

This can therefore be written as the product of the z-transforms of h(k) and x(n) as in Equation 6.104.

$$Y(z) = H(z) X(z) \qquad (6.104)$$

Other important properties of the z-transform are shown in Table 6.2.

In both signal processing applications and linear systems theory, linear recursive operators are used extensively, examples in both areas being the design of recursive digital filters amd feedback systems in control, respectively, and the general input-output relationship for such a linear recursive system can be written in the form shown in Equation 6.105.

$$\sum_{k=0}^{p} a_k y(n-k) = \sum_{k=0}^{q} b_k x(n-k) \qquad (6.105)$$

Taking the z-transform of both sides of the expression, and interchanging the order of summation we obtain Equation 6.106.

$$\sum_{k=0}^{p} a_k \left[\sum_{n=-\infty}^{\infty} y(n-k) z^{-n} \right] = \sum_{k=0}^{q} b_k \left[\sum_{n=-\infty}^{\infty} x(n-k) z^{-n} \right] \qquad (6.106)$$

Using the shift theorem, this expression can be written as in Equation 6.107.

Fourier analysis

Table 6.2 Properties of the z-transform

Property	Time series	z–transform	Region of convergence
	$x(n)$	$X(z)$	$r_{cx} < |z| < r_{ax}$
	$y(n)$	$Y(z)$	$r_{cy} < |z| < r_{ay}$
Linearity	$ax(n) + by(n)$	$aX(z) + bY(z)$	At least max (r_{cx}, r_{cy}) $< |z| <$ min (r_{ax}, r_{ay})
Time shift	$x(n-m)$	$z^{-m} X(z)$	$r_{cx} < |z| < r_{ax}$
Convolution	$\sum_{k=-\infty}^{\infty} x(k) y(n-k)$	$X(z) Y(z)$	At least max (r_{cx}, r_{cy}) $< |z| <$ min (r_{ax}, r_{ay})
Exponential multiplication	$a_n x(n)$	$X(a^{-1} z)$	$|a| r_{cx} < |z| < |a| r_{ax}$
Time multiplication	$nx(n)$	$-z \dfrac{dX(z)}{dz}$	$r_{cx} < |z| < r_{ax}$
Product	$x(n) y(n)$	$\dfrac{1}{2\pi j} \oint_C X(w) \times Y\left(\dfrac{z}{w}\right) w^{-1} dw$	$r_{xc} r_{cy} < |z| < r_{ax} r_{ay}$
Correlation	$\sum_{k=-\infty}^{\infty} x(k) y(n+k)$	$X(z^{-1}) Y(z)$	max $(r_{ax}^{-1}, r_{cy}) < |z|$ $<$ min (r_{cx}^{-1}, r_{ay})
Time transpose	$x(-n)$	$X(z^{-1})$	$r_a^{-1} < |z| < r_c^{-1}$

$$\left[\sum_{k=0}^{p} a_k z^{-k}\right] Y(z) = \left[\sum_{k=0}^{q} b_k z^{-k}\right] X(z) \tag{6.107}$$

This is used to define the transfer function, $H(z)$, of the system, which may be written as in Equation 6.108 where $H(z)$ is given by Equation 6.109.

$$Y(z) = H(z) X(z) \tag{6.108}$$

112 Linear time-invariant digital systems

$$H(z) = \frac{b_0 + b_1 z^{-1} + \ldots + b_q z^{-q}}{a_0 + a_1 z^{-1} + \ldots + a_p z^{-p}} \tag{6.109}$$

From the transfer function of a system, the frequency response is found by evaluating the transfer function at $z = e^{j\omega}$, where the angular frequency $\omega = 2\pi f$.

Factorising the above transfer function as in Equation 6.110, where the zeros and poles, z_i and p_i, may be complex numbers, and evaluating this function at $z = e^{j\omega}$, we obtain the associated frequency response as in Equations 6.111 and 6.112.

$$H(z) = \frac{b(z-z_1)(z-z_2)\ldots(z-z_q)}{(z-p_1)(z-p_2)\ldots(z-p_p)} \tag{6.110}$$

$$H(e^{j\omega}) = \frac{b(e^{j\omega}-z_1)(e^{j\omega}-z_2)\ldots(e^{j\omega}-z_q)}{(e^{j\omega}-p_1)(e^{j\omega}-p_2)\ldots(e^{j\omega}-p_p)} \tag{6.111}$$

$$H(e^{j\omega}) = \frac{b\alpha_1(\omega)\alpha_2(\omega)\ldots\alpha_q(\omega)}{\beta_1(\omega)\beta_2(\omega)\ldots\beta_p(\omega)} \tag{6.112}$$

The complex functions $\alpha_i(\omega)$ and $\beta_i(\omega)$ are given by Equations 6.113 and 6.114.

$$\alpha_i(\omega) = e^{j\omega} - z_i \tag{6.113}$$

$$\beta_i(\omega) = e^{j\omega} - p_i \tag{6.114}$$

A vector interpretation of the frequency response for a linear zeros system with three poles and two zeros is shown in Figure 6.10 and Equations 6.115 and 6.116.

$$|H(e^{j\omega})| = \frac{|b||\alpha_1|*|\alpha_2|}{|\beta_1|*|\beta_2|*|\beta_3|} \tag{6.115}$$

angle $\{H(e^{j\omega})\} = \Theta_1 + \Theta_2 - \Phi_1$
$- \Phi_2 - \Phi_3 +$ angle $\{b\}$ (6.116)

6.9 The inverse z-transform

In the section describing the Fourier transform, it was straightforward to relate $x(n)$ to $X(k)$ through the definition of the integral transform, and vice versa. In this section we will consider methods of obtaining the inverse z-transform, that is, to obtain $x(nT)$ given $X(z)$.

The formal method is derived using Cauchy's integral theorem. From the definition of the z-transform, multiplying both sides by z^{m-1} and integrating around a closed contour in the z-plane, gives Equation 6.117.

$$\oint_C X(z) z^{m-1} dz = \oint_C \sum_{n=0}^{\infty} x(nT) z^{m-n-1} dz \qquad (6.117)$$

If the path of integration encloses the origin, then from Cauchy's integral theorem, the right hand side is zero except when $m = n$, in which case it is $2\pi j$.

Therefore Equation 6.118 can be obtained, which may be evaluated using the residue theorem.

$$x(nT) = \frac{1}{2\pi j} \oint_C X(z) z^{n-1} dz \qquad (6.118)$$

The remaining methods involve partial fraction expansions and straightforward inversion by division.

6.10 Data truncation

As mentioned before, the Fourier transform of any signal satisfies the periodic relationship as in Equation 6.119.

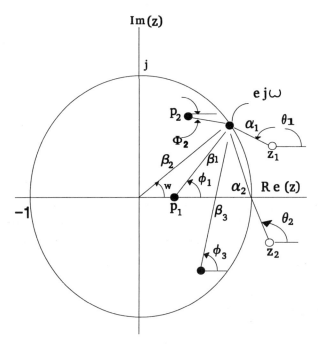

Figure 6.10 Geometrical construction of the frequency response for a linear system with three poles and two zeroes

$$X\left(e^{j(\omega+2\pi)}\right) = X\left(e^{j\omega}\right) \tag{6.119}$$

Due to this periodicity, one can consider the frequency range $-\pi < \omega < \pi$ to be the fundamental set of frequencies.

As seen before, the Fourier transform is of fundamental importance in characterising the spectral content of signals, be they in the time domain, spatial domain, etc. However, the Fourier transform requires the entire set of the signal samples, and the question therefore arises concerning the applicability of transform techniques when the data are of finite support.

The procedure of filling in the 'missing data' with zeros is a very common method of trying to account for the situation.

As before, the discrete Fourier transform of a finite data set can be written as in Equation 6.120.

Fourier analysis 115

$$X_N(e^{j\omega}) = \sum_{n=0}^{N-1} x(n) e^{-j\omega n} \quad (6.120)$$

Replacing $x(n)$ by its inverse Fourier transform gives Equation 6.121.

$$X_N(e^{j\omega}) = \sum_{n=0}^{N-1} \left[\frac{1}{2\pi} \int_{-\pi}^{\pi} X(e^{jv}) e^{jv} dv \right] e^{-j\omega n} \quad (6.121)$$

Interchanging the order of the integral and the summation, and using the definition of Equation 6.122 gives Equation 6.123, where * means convolution of the non-truncated Fourier transform $X(e^{j\omega})$ with the rectangular window transform $W_r(e^{j\omega})$.

$$W_r(e^{j\omega}) = \sum_{n=0}^{N-1} e^{-j\omega n}$$
$$= e^{-j\omega(N-1)/2} \frac{\sin(\omega N/2)}{\sin(\omega/2)} \quad (6.122)$$

$$X_n(e^{j\omega}) = \frac{1}{2\pi} \int_{-\pi}^{\pi} X(e^{jv}) W_r(e^{j(\omega-v)}) dv$$
$$= \frac{1}{2\pi} X(e^{j\omega}) * W_r(e^{j\omega}) \quad (6.123)$$

When only a finite amount of data is present, the effect in trying to estimate the underlying Fourier transform manifests in terms of the convolution of the underlying transform with the transform of a rectangular window. This means that in the time domain, the data have been 'multiplied' by a window which is unity over the duration of the data, and zero outside. As the amount of data increases, the Dirichlet kernel, W_r, tends towards a Dirac delta function. The Dirichlet kernel is plotted as a function of frequency, ω, in Figure 6.11, and it can be seen to consist of many lobes centred at frequencies 0, and

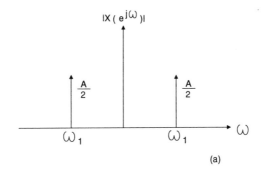

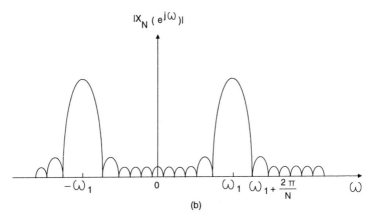

Figure 6.11 Magnitude of the Fourier transform associated with: (a) an untruncated sinusoid; (b) a truncated sinusoidal time series

$\frac{k\pi}{N}$ for $k = 3, 5, 7, \ldots$ The main lobe has a width of $\frac{4\pi}{N}$ and an amplitude of N, and the side lobes each have a width of $\frac{2\pi}{N}$.

To determine the effects of using only a finite amount of data, consider the example where the data consist of a pure sinusoid as in Equation 6.124.

Fourier analysis

$$x(n) = A \cos(n\omega_1) \qquad (6.124)$$

The corresponding Fourier transform is given by Equation 6.125 for $-\pi < \omega \leq \pi$.

$$X(e^{j\omega}) = \pi A [\delta(\omega - \omega_1) + \delta(\omega + \omega_1)] \qquad (6.125)$$

The Fourier transform of the truncated sequence is therefore given by Equation 6.126.

$$X_N(e^{j\omega}) = A e^{-j(\omega-\omega_1)(N-1)/2} \frac{\sin\left[\frac{(\omega-\omega_1)N}{2}\right]}{\sin\left[\frac{(\omega-\omega_1)}{2}\right]}$$

$$+ A e^{j(\omega+\omega_1)(N-1)/2} \frac{\sin\left[\frac{(\omega+\omega_1)N}{2}\right]}{\sin\left[\frac{(\omega+\omega_1)}{2}\right]} \qquad (6.126)$$

Thus the effect of using only a small data set is to smear out the original delta function spectrum. This effect is commonly known as spectral leakage.

In attempts to reduce this spectral leakage, a variety of so called data windows have been introduced in order to 'smooth out' the ripples introduced by the rectangular window. A few of the window types are, triangular, Hanning (or raised cosine), Hamming, Blackman, etc. For a good introduction to windowing see Harris (1978). The effects of various data windows are shown in Table 6.3.

Clearly there is a trade off between side lobe levels and width of the main peak, so caution is needed in using data windows depending on the application at hand. One should also say that there is a school of thought that windowing the data by anything other than a rectangular window distorts the original data, and depending on the application, (spectral analysis, for example), maybe model based signal

Table 6.3 Commonly used data windows

Window	$w(n), 0 \leq n \leq (N-1)$	Main-lobe width (rad)	20 log (main-lobe amplitude)/(Largest side-lobe amplitude)
Rectangular	1	$\dfrac{4\pi}{N}$	-13dB
Bartlett	$\dfrac{2n}{N-1} \quad 0 \leq n \leq \dfrac{N-1}{2}$ $2 - \dfrac{2n}{N-1} \quad \dfrac{N-1}{2} < n \, N-1$	$\dfrac{8\pi}{N}$	-27dB
Hanning	$0.5 \left[1 - \cos\left(\dfrac{2\pi n}{N} \right) \right]$	$\dfrac{8\pi}{N}$	-32dB
Hamming	$0.54 - 0.46 \cos\left(\dfrac{2\pi n}{N} \right)$	$\dfrac{8\pi}{N}$	-43dB
Blackman	$0.42 - 0.5 \cos\left(\dfrac{2\pi n}{N} \right)$ $+ 0.8 \cos\left(\dfrac{4\pi n}{N} \right)$	$\dfrac{12\pi}{N}$	-58dB

processing is more appropriate than just calculating a window based FFT; but again, this will depend on the precise nature of the application (Jaynes, 1987).

6.11 Conclusions

In this chapter, we have introduced the concept of orthogonal basis functions and the methods of obtaining the generalised Fourier coefficients which can be used in order to expand a particular signal of interest in terms of 'elementary' basis functions, for example sines and cosines.

Consideration was given to the representation of continuous signals and functions, and the analysis then considered the case of discrete data sequences.

The Fourier transform was introduced and applied to aperiodic signals and the relationship between the power in a signal and the expansion coefficients was given. Ideas concerning spectral analysis were introduced.

Linear systems theory was briefly touched upon, the z-transform was introduced and the concepts of transfer function and frequency response were discussed.

Fast implementations of the discrete Fourier transform, using the FFT, were discussed, and the ideas of spectral leakage and windowing were mentioned.

There exists a vast literature concerned with function expansion and representation, some references for which are given below.

The subject of transform methods is similarly vast, and it must be appreciated that the Fourier and z-transforms are members of a transform family including radon transforms, Wigner transforms, wavelet transforms and Hartley transforms, (Bracewell,1986).

6.12 References

Banks, S. (1990) *Signal Processing, Image Processing and Pattern Recognition*, Prentice Hall.

Bracewell, R.N. (1986) *The Hartley Transform*, Oxford University Press.

Burrus, C.S., Parks, T.W. (1985) *DFT/FFT and Convolution Algorithms*, Wiley Interscience.

Harris, F.J. (1978) *Proc. 66, Number 1 On the use of windows for harmonic analysis with the discrete Fourier transform*, IEEE.

Jaynes, E.T. (1987) Bayesian spectrum and Chirp analysis. *In Maximum Entropy and Bayesian Spectral Analysis and Estimation Problems*, Kluwer.

Kamen, E.W. (1990) *Introduction to Signals and Systems*, Maxwell Macmillan.

Walker, J.S. (1988) *Fourier Analysis*, Oxford University Press.
Wax, N. (ed.) (1954) *Noise and Stochastic Processes*, Dover.

7. Queuing theory

7.1 Introduction

7.1.1 Some queuing problems

Queuing is a ubiquitous feature of electronic systems such as computers and communication networks. Queues will build up, even if the capacity of the system exceeds the load on it. This is an inevitable consequence of the fact that arrivals and services take place in a random fashion. Indeed the more random the services and the arrivals are, the more queues build up. Thus it is possible for long queues to develop at even quite light loads arising as a consequence of sheer variability in the pattern of arrivals and services.

If the queue length distribution does not depend on time the queue is said to be in equilibrium. This is only possible if the queue's capacity is not exceeded. In most queuing situations the equilibrium builds up rapidly. Even when the load itself is not constant the changes in it take place more slowly than in the queue so that the equilibrium is effectively 'tracked'. This means that equilibrium models can be used to determine all the usual key performance measures: delay, queue lengths, etc. Based on this, the results here are confined to equilibrium models.

Typical delay performance for a single server queue in equilibrium with a single customer class is shown in the steadily increasing graph depicted in Figure 7.1. The horizontal axis is the occupancy (arrival rate/service rate), which is the proportion of time the customers keep the server busy. The vertical axis is average delay. As the graph shows, there is an increasing delay penalty for using the system closer and closer to full capacity (occupancy =1). Also shown is the graph for a deterministic system (constant time between arrivals and constant service time). No queue builds up until the system is at full load.

122 Introduction

Figure 7.1 Delay performance of a single server queue

When the occupancy is less than 1 it is said that the queue is stable and an equilibrium will then exist, as already mentioned.

In most queuing situations arrivals and services do not take place in a deterministic fashion and so it is the former graph which is more representative of system performance. A fast response is a strict requirement for many electronic systems and so modelling delays and finding strategies which make efficient use of the resources within these systems becomes an essential part of their design. For example if the graph represents the performance of a computer controlling a telephone exchange, one way to reduce the average delays would be to schedule the work so as to process shorter jobs, such as call set-ups and clear downs first. This procedure 'shares out' delay in a sensible fashion. Longer jobs, for example those associated with background routines are given lower priority and are held back. Telephone calls are connected as quickly as possible, whilst the longer jobs take on an increase in delay which is often small in comparison with the amount of time needed to process them.

It is not always quite so obvious what the capacity of a queuing system is as it was in the example depicted in Figure 7.1, where the queue remains stable provided the arrival rate is less than the service rate. A communication channel which allows a group of users to share it by means of random multiple access protocols, is a case in point. Such a channel will be occupied not only with the transmission of users' information but also with reservation periods where the users actually obtain rights to transmit. Some time is used in transmissions, some in reservation but how much time is spent in each task? This depends on the nature of the protocols themselves.

For example, consider the operation of slotted ALOHA depicted in Figure 7.2. Packets of information are all the same length and a packet is received successfully if it is the only one to be transmitted in a particular slot. If two or more are transmitted there is a collision and none are successfully received, such packets are retransmitted at random in later slots. But there are many ways to choose a slot for retransmission at random and this choice is critical. It determines how successful the protocol is in avoiding collisions and how many packets can be eventually transmitted. A poor protocol choice may give very little useful throughput, or, worse still, cause the channel to jam. In this case it turns out that using fixed retransmission probabilities does not work very well. It makes no allowance for the variation in the backlog of packets awaiting retransmission. However, adjusting the retransmission probability according to an estimate of this backlog does perform adequately. Each packet is retransmitted with high probability when the backlog is estimated to be low and with low probability when the estimated backlog is high.

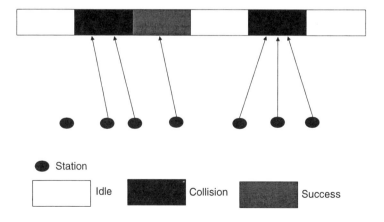

Figure 7.2 The slotted ALOHA protocol

In queuing network problems there is more than one queue and customers make a series of visits from one queue to the next. In such networks there are interactions between the individual queues within it. For example the pattern of arrivals at one queue is determined by the departure patterns from the queues which have customers feeding into it (together with the new arrivals to the network). Since these departure patterns themselves depend on the arrivals to those queues analysis can prove very difficult.

It is interesting to learn, therefore, that for one very wide class of networks it is relatively easy to determine the equilibrium distribution of the customers within the network. Furthermore the form of this distribution is particularly striking. It is simply the product of the equilibrium distribution for each queue operating as if it were in isolation!

Such models have been used to analyse the performance of packet switched networks. In these networks, packets of information are transmitted from source to destination via a series of transmission links. These transmission links are interconnected by switching nodes which queue the packets in storage buffers until it is their turn to be transmitted. (Packet switched networks are sometimes referred to as store and forward networks.)

From these models it is known that the delays which packets experience and, indeed, the throughput of the network itself depends strongly on the routes which the packets take through it. But determining the best choice of routes using such models is not easy because the assignment of one particular route or set of routes to one group of packets has an effect on the delays which other packets experience. However effective routing algorithms have been proposed based on models such as the ones mentioned above (Gallager, 1977).

7.1.2 Little's law

There are very few completely general queuing results. However there is one, which is of considerable utility for the analysis of queues in equilibrium, Little's law. It even applies to systems which can hardly be thought of as queues at all. Suppose that customers arrive

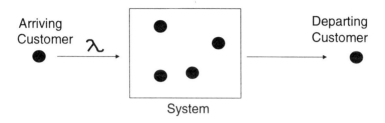

Figure 7.3 Illustration of Little's law

to the system (depicted as a black box in Figure 7.3) at a rate λ and that the average sojourn time for each customer is $E[D]$, in equilibrium. Suppose further that the time average number of customers within the system is $E[N]$. Then Little's law gives Equation 7.1.

$$\lambda E[D] = E[N] \qquad (7.1)$$

Little's law is an accounting identity and can be understood as such. Suppose customers are charged at a rate of £1 per second they remain in the system. There are two equivalent ways in which the money can be collected. The first is to collect the appropriate amount from each customer as they leave. Since the average duration within the system is $E[D]$ and the mean departure rate must be λ the rate at which money is being collected is $\lambda E[D]$. The other way to collect the money is for the customers to pay continuously as they remain in the system. Since the average number of customers within the system is $E[N]$ the rate at which money is being collected is also $E[N]$. By altering the rates of payment one may obtain other similar identities.

7.1.3 Kendall queuing notation

As randomness plays such an important part in queuing performance queues are categorised accordingly. The notation A/B/m/S/P is in common use, as in Table 7.1.

126 Introduction

Table 7.1 Notations in common use

Symbol	Item
A	Interarrival distribution
B	Service distribution
m	Number of servers
S	Storage capacity
P	Customer population

Table 7.2 Distribution symbols in use with Kendall queuingnotation

Item	Symbol
Deterministic	D
Erlangian with k degrees of freedom	E_k
General	G
Hyperexponential with j degrees of freedom	H_j
Markovian	M

Omission of either the storage or the population entry means they are infinite. The most often used interarrival and service distributions are given in Table 7.2.

Markovian is equivalent to a Poisson process, if the customer population is infinite, and to exponential for the service distribution. The Erlangian distribution is just the sum of k independent, identically distributed exponential variates. Sometimes this is referred to as a server having k exponential service 'phases'. See Figure 7.4(a) illustrates E_3. (This distribution is also referred to as the gamma.) The hyperexponential is constructed from the exponential distribution as

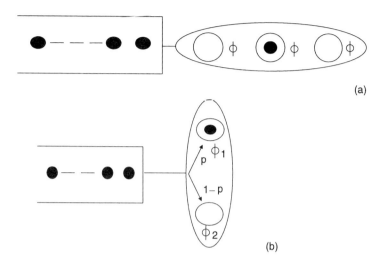

Figure 7.4 An Erlangian E_3 and a Hyperexponential H_2 server: (a) E_3; (b) H_2

well but the customer undergoes a single service phase allocated at random from a number of possible phases, see Figure 7.4(b).

Some examples of queues represented in Kendall queuing notation are:

1. D/G/1; single server queue with deterministic arrivals and an arbitrary service time.
2. M/M/3/10; a queue with Poisson arrivals, three exponential servers and room for 10 customers, including those in service.
3. M/M/1//20; a queue with exponential interarrival times with rate proportional to the number of customers not in queue, a single exponential server, and infinite waiting room. The customer population is 20.

In the case of queuing networks one may wish to specify the queue types whilst ignoring the nature of the arrival process to it altogether. In such cases it may be written, for example, that there is a network

7.2 Models

7.2.1 The memoryless property and the Poisson process

The exponential distribution is encountered frequently in queuing analysis. One reason is that the exponential can be used as a building block to construct other distributions as has been shown earlier. Indeed the distribution of virtually any positive random variable may be approximated using the exponential (Kelly, 1979). However, the following property is more significant reason for its importance in queuing theory.

The density of the exponential is $\varphi e^{-\varphi s}$ and integration of this gives the corresponding distribution function $1 - e^{-\varphi s}$. Suppose X is the random variable drawn from this distribution. Say that X, "completes" when the time corresponding to X is reached starting from 0. Suppose t seconds have elapsed and X has not completed. The memoryless property is this: that the distribution of remaining time until X completes in no way depends on t and is given by the same exponential distribution as X.

To see this must be the case, consider the following example. A random job from a class with exponential processing time requirements, with rate parameter φ, has been the sole task within a computer for t seconds. The total processing requirement is X. Let us compute the probability that the job will take at most a further s seconds to be completed. The proportion of random jobs whose processing times lie in the interval $(t,t+s)$ is given by the area A under the graph in Figure 7.5, as marked. However this particular job is one which has taken t seconds already and it is the proportion of all such jobs which are completed within a further s seconds that is required. The probability a job takes at least t seconds is given by the tail region B. Thus the conditional distribution of the remaining time until the job is completed, given that t seconds have elapsed, is as in Equation 7.2.

Queuing theory 129

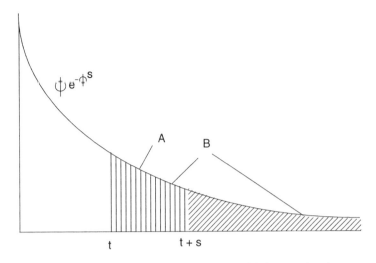

Figure 7.5 Conditional probability that the job is completed within s seconds

$$\Pr\{X \leq t+s \mid t < X\} = \frac{\text{Area } A}{\text{Area } B}$$

$$= \frac{\Pr\{t < X \leq t+s\}}{\Pr\{t < X\}}$$

$$= \frac{e^{-\varphi t} - e^{-\varphi(t+s)}}{e^{-\varphi t}} = 1 - e^{-\varphi s} \qquad (7.2)$$

This is seen to be independent of t. The conditional probability that the job is completed within a further s seconds is given by the very same exponential distribution as would be used to determine the probability of the job being completed in s seconds starting from time 0.

Often it is said that the job is being processed at rate φ, meaning that the probability that the job is completed in the next infinitesimal time interval dt is $\varphi \, dt$, given it has not been completed already.

The Poisson process may be constructed from the exponential. Indeed, it is useful to regard the Poisson process as a sequence of customers arriving at a queue with mutually independent and identical exponential interarrival times. As such, the Poisson process 'inherits' the memoryless property from the exponential. This means that the probability an arrival takes place in any given interval is independent of the history of the queue prior to that interval and in particular of the state of the queue immediately before it arrives. The parameter of the Poisson process is the same rate parameter as the underlying exponential distribution.

In fact given a sequence of independent random variable, $X_1, X_2, X_3, \ldots$ with common exponential distribution rate parameter λ, the sequence of arrival times is determined by expression 7.3, starting from time 0.

$$X_1, X_1 + X_2, X_1 + X_2 + X_3, \ldots \qquad (7.3)$$

Now that the construction follows, the Poisson distribution can be checked. To do this recall that the density of the sum of k independent exponential random variable with rate parameter λ is given by expression 7.4, which is the k-Erlangian distribution with parameter λ (Feller, 1970).

$$\lambda \frac{(\lambda t)^{k-1}}{(k-1)!} e^{-\lambda t} \qquad (7.4)$$

Denote the number of arrivals in time interval $(0,t]$ by N_t, then Equation 7.5 may be obtained.

$$\Pr\{N_t = k\} = \Pr\{X_1 + \ldots + X_k \leq t; X_1 + \ldots + X_k + X_{k+1} > t\}$$

$$= \int_0^t \lambda \frac{(\lambda s)^{k-1}}{(k-1)!} e^{-\lambda s} e^{-\lambda(t-s)} \, ds$$

$$= \frac{(\lambda t)^k}{k!} e^{-\lambda t} \qquad (7.5)$$

7.2.2 Markov processes and Markov chains

Markov process models underlie a great deal of queuing analysis and a grasp of these models is essential to understanding analytical models of queues. A stochastic process $X(s)$ defined on a countable state space K is said to be Markov if, given a finite sequence of $n > 1$ points (nearly always points in time) $s_1 < s_2 ... < s_n$, Equation 7.6 holds.

$$\Pr\{X(s_n) = j_n \mid X(s_1) = j_1; ....; X(s_{n-1}) = j_{n-1}\}$$
$$= \Pr\{X(s_n) = j_n \mid X(s_{n-1}) = j_{n-1}\}, j_1, ... j_n \; \varepsilon \; K \quad (7.6)$$

The conditioning event must have positive probability. In words, the probability of finishing up in a particular state conditional on the process visiting given states at given times depends only on the most recent of those states stipulated. Thus to determine the probability of future events one need only take the current state into account. Notice that the above probability, the transition probability from j_{n-1} to j_n, may also depend on the time, but if it depends only on the difference $s_n - s_{n-1}$ the process is called time homogeneous. If any state can be attained from any other state the process is said to be irreducible.

A discrete time Markov process is called a Markov chain. For simplicity it may be supposed that transitions take place at unit intervals. In Markov chains it is possible that certain states can be visited at intervals which are constant multiples of some given number $v > 1$. This is illustrated in Figure 7.6 which is for the absolute difference of heads to tails in a series of coin tossings. The even states are visited every other transition, so $v = 2$. Such a chain is called periodic. We shall not consider such chains here. All the chains considered will be aperiodic.

Time homogeneous Markov chains which are irreducible and aperiodic, may have an equilibrium probability distribution $\pi(j)$ given by Equation 7.7, where $p(k,j)$ is given by Equation 7.8.

$$\pi(j) = \sum_{k \, \varepsilon \, K} \pi(k) p(k,j) \quad (7.7)$$

132 Models

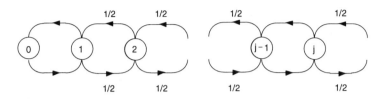

Figure 7.6 A coin tossing chain with period 2

$$p(k,j) = \Pr\{X(s+1) = j | X(s) = k\} \tag{7.8}$$

If an equilibrium distribution exists the distribution of the process will always converge to it as in Equation 7.9, and, having this property, it must be unique. $pi(j)$ can be regarded as the long run average proportion of time the process spends in state j.

$$\lim_{s \to \infty} \Pr\{X(s) = j | X(0) = k\} = \pi(j) \tag{7.9}$$

As far as the Markov processes here are concerned, they may be regarded as arising in the following way. The construction is based on the exponential which was discussed in the previous section. In each state j the rate at which transitions are made to state k is $q(j,k)$. Thus the probability that the process enters state k in the next interval of length dt is $q(j,k)dt$. The probability that a transition out of state j will take place at all in the time interval dt is therefore given by Equation 7.10.

$$\sum_{k \neq j} q(j,k) \, dt = q(j) \, dt \tag{7.10}$$

Put another way, each visit to j has a duration which is exponential with rate parameter $q(j)$. The probability that the process then enters state k is given by Equation 7.11.

$$p(j,k) = \frac{q(j,k)}{q(j)} \tag{7.11}$$

The equilibrium distribution, if it exists, satisfies Equation 7.12, which are referred to as the global balance equations. $\pi(j) q(j,k)$ is called the probability flux out of j into state k in equilibrium.

$$\sum_{k \varepsilon K} \pi(j) q(j,k) = \sum_{k \varepsilon K} \pi(k) q(k,j) \qquad (7.12)$$

7.2.3 Birth-Death models of queues

The underlying idea of this class of queuing models is to regard arriving customers as births and departures as deaths thus the queue alters by at most one. The most general form of linear birth-death process is depicted in Figure 7.7.

The global balance equations can be obtained by equating the probability flux out of each state with the probability flux into it as already described. (Equation 7.13.)

$$\lambda_0 \pi_0 = \mu_1 \pi_1 ;$$

$$(\lambda_k + \mu_k) \pi_k = \lambda_{k-1} \pi_{k-1} + \mu_{k+1} \pi_{k+1}, k \geq 1 \qquad (7.13)$$

However these equations may be solved by using a simpler set of equations, known as the local (detailed) balance equations, as in Equation (7.14).

$$\lambda_k \pi_k = \mu_{k+1} \pi_{k+1}, k \geq 0 \qquad (7.14)$$

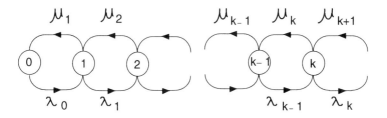

Figure 7.7 A linear birth-death model

134 Models

From these equations the equilibrium solution is easily seen to be as in Equation 7.15, and π_0 is found from the fact that the probabilities must sum to 1.

$$\pi_k = \pi_0 \prod_{j=0}^{j=k} \frac{\lambda_j}{\mu_{j+1}} \ , k = \sim 1 \tag{7.15}$$

It can be seen that if a probability distribution which satisfies the detailed balance equations can be found then it must satisfy the global balance equations. This is so because each global balance equation is the sum of two detailed balance equations, excluding the first and the last, if there is a last. These two are just a single local balance equation. (Apart from the simplification provided, the significance of the local balance equations is that they show that the linear birth death process is reversible in equilibrium. In crude terms, reversible Markov processes cannot be distinguished, statistically, going forwards in time from going backwards in time. From this it is clear that a Markov process must be in equilibrium to be reversible.)

By taking different birth and death rates one can obtain a wide range of queuing models, for example as in Table 7.3.

Note that in each case the mean sojourn time, $E[D]$, can be found by an application of Little's law. However in the case of M/M/K//L there is only a finite number of customers and in the case of M/M/1/S, customers are lost because there is only finite storage. In neither case is the arrival rate of customers entering queue λ.

For M/M/K//L Equation 7.16 may be obtained, where θ is the mean rate at which the customers complete a cycle of standing idle, queuing and service.

$$\theta \left(E[D] + \frac{1}{\lambda} \right) = L, E[Q] = \theta E[D] ; \tag{7.16}$$

Equation 7.16 follows from Little's law applied to all the customers, $\frac{1}{\lambda}$ is the mean time a customer stands idle before returning to

Table 7.3 Equilibrium queue distribution for some birth-death models

Queue type	Parameters	Equilibrium queue distribution
M/M/1	$\lambda_k = \lambda, k \geq 0$ $\mu_k = \mu, k \geq 1$	$Q_k = (1-\rho)\rho^k, k \geq 0$ $E[Q] = \dfrac{\rho}{(1-\rho)}$
M/M/K//L	$L > K, \lambda_k$ $= (L-k)\lambda$ $\mu_k = k\mu, k \leq K$ $\mu_k = K\mu, k > K$	$Q_k = \binom{L}{k}\left(\dfrac{\lambda}{\mu}\right)^k Q_0, k \leq K$ $Q_k = \binom{L}{k}\left(\dfrac{\lambda}{\mu}\right)^k \dfrac{k!}{K!} K^{K-k} Q_0$ $K < k \leq L$
M/M/1/S	$\lambda_k = \lambda, k < S$ $\lambda_k = 0, k \geq S$ $\mu_k = \mu, k > 0$	$Q_k = \dfrac{(1-\rho)}{\left(1-\rho^{S+1}\right)}\rho^k, k \leq S$ $E[Q] = \dfrac{\rho}{(1-\rho)} - \dfrac{(S+1)\rho^{S+1}}{1-\rho^{S+1}}$
M/M/∞	$\lambda_k = \lambda$, $\mu_k = k\mu$	$Q_k = e^{-\rho}\dfrac{\rho^k}{k!}, k \geq 0$ $E[Q] = \rho$

queue for service. The second part comes from Little's law applied to the queuing and service stages, $\sum_k k Q_k = E[Q]$.

Substituting for θ gives Equations 7.17 and 7.18.

$$E[Q] = \frac{LE[D]}{\left(E[D] + \dfrac{1}{\lambda}\right)} \tag{7.17}$$

$$E[D] = \frac{E[Q]}{\lambda(L - E[Q])} \tag{7.18}$$

For M/M/1/S the mean arrival rate of customers entering queue is given by expression 7.19.

$$\frac{\lambda(1-\rho^S)}{(1-\rho^{S+1})} \tag{7.19}$$

Again from Little's law we have Equation 7.20.

$$\begin{aligned}
E[D] &= \left[\frac{\rho(1-\rho^{S+1})}{(1-\rho)(1-\rho^S)} - \frac{(S+1)\rho^{S+1}}{(1-\rho^S)} \right] \frac{1}{\lambda} \\
&= \left[\frac{(1-\rho^{S+1})}{(1-\rho)(1-\rho^S)} - \frac{(S+1)\rho^S}{(1-\rho^S)} \right] \frac{1}{\mu}
\end{aligned} \tag{7.20}$$

7.3 Transform methods

7.3.1 Arival and departure distributions

It is possible to think of the distribution of the number of customers in a queue in three ways:

1. The distribution of the number of customers found in queue by an arriving customer.
2. The distribution of the number of customers left in queue by a departing customer.
3. The distribution of the number of customers in queue at a particular time.

Fortunately it is the case that, in equilibrium, the first two of these distributions coincide in very general circumstances. This will be demonstrated in a moment. It is even possible that all three of these

Queuing theory

distributions will coincide in equilibrium, this is the case when the arrival process is Poisson.

For the equality of the first two distributions in equilibrium it is sufficient that arrivals and departures should take place singly and that at least one of the two equilibrium distributions should exist. This latter condition holds for most queues, even non-Markovian ones, provided that the queue is stable. To see this suppose, for simplicity, that the queue is initially empty. (The argument also works when there is an initial queue of customers but this minor complication is ignored.)

Figure 7.8 shows the kth departure leaving $j \leq m$ customers still in the system. This means that customer $k+m+1$ is yet to arrive and that when he does he will find m or less customers in the system. (Equation 7.21.)

$$\Pr\{d_k \leq m\} \geq \Pr\{a_{m+k+1} \leq m\} \qquad (7.21)$$

Now consider the arrival of customer $m+k+1$ and suppose that he too finds $j \leq m$ customers in the system. This means that the kth departure has already taken place and left behind at most m customers. (Equation 7.22.)

$$\Pr\{a_{k+m+1} \leq m\} \geq \Pr\{d_k \leq m\} \qquad (7.22)$$

Hence Equation 7.23 can be obtained, and if we suppose that the equilibrium arrival distribution $A(m)$ exists then Equation 7.24 follows.

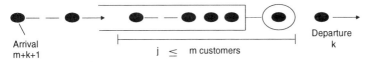

Figure 7.8 Customers 'seen' on arrival and departure

$$\Pr\{a_{k+m+1} \le m\} = \Pr\{d_k \le m\} \tag{7.23}$$

$$A(m) = \lim_{k \to \infty} \Pr\{a_{k+m+1} \le m\}$$

$$= \lim_{k \to \infty} \Pr\{d_k \le m\} = D(m) \tag{7.24}$$

Similarly the equilibrium departure distribution can be shown to exist provided the equilibrium arrival distribution does and the two are once again equal.

7.3.2 Outline of the method

The approach is based on a transformation of the Markov equations which describe the queuing process and so obtaining algebraic equations for the transforms, and then solving these equations. This latter step usually involves finding zeros of functions. The method can be applied to both equilibrium and transient queues (Takacs, 1962).

The most commonly used transforms are the probability generating function (z-transform), for discrete random variables and the Laplace transform for non-negative continuous random variables. If N is a discrete random variable and T a non-negative continuous (or mixed) random variable, Equation 7.25 may be obtained, where p_k is given by Equation 7.26 and $H(t)$ is the distribution of T.

$$\begin{aligned} f(z) &= E[z^N] = \sum_{k=0}^{\infty} p_k z^k, H^*(s) \\ &= E[e^{-sT}] = \int_0^{\infty} e^{-st} dH(t) \end{aligned} \tag{7.25}$$

$$p_k = \Pr\{N = k\} \tag{7.26}$$

A little care is needed in working with the Laplace transforms of queuing variables since mixed distributions can occur, e.g. waiting times which are a mixture of an atom at the origin (no wait at all) and a continuous density (waiting time conditional on having to wait).

Moments can be obtained from these by taking derivatives at 1 and 0 respectively, as in Equation 7.27. The derivatives give factorial moments in the case of the probability generating functions and moments about the origin (to within a sign) in the case of the Laplace transform.

$$\frac{d^n f}{dz^n} = E[N(N-1)\ldots(N-n+1)], \frac{d^k H^*}{ds^k}$$
$$= (-1)^k E[T^k] = (-1)^k h_k \tag{7.27}$$

7.3.3 The M/G/1 queue in equilibrium

In general the M/G/1 queue is not Markov unless one includes the time to complete the customer in service as well as the number of customers in the queue.

However an analysis can be conducted by examining the queue at special instants. In fact the number of customers in queue between each departure is a Markov chain (referred to as the embedded Markov chain).

This is so because the arrival process is Poisson.

Suppose the arrival rate of customers is ν and the service distribution is $H(x)$ with density $h(x)$.

The probability p_k that k customers arrive during a customer service is given by Equation 7.28.

$$p_k = \int_0^\infty e^{-\nu x} \frac{(\nu x)^k}{k!} h(x) \, dx \tag{7.28}$$

Transforming to get the generating function for the p_k's gives Equation 7.29.

140 Transform methods

$$\sum_{k=0}^{\infty} p_k z^k = \int_0^{\infty} e^{-\upsilon x} \frac{(\upsilon z x)^k}{k!} h(x) \, dx$$

$$= \int_0^{\infty} e^{-\upsilon x} \sum_{k=0}^{\infty} \frac{(\upsilon z x)^k}{k!} h(x) \, dx$$

$$= \int_0^{\infty} e^{-\upsilon x + \upsilon x z} h(x) \, dx = H^*(\upsilon - \upsilon z) \quad (7.29)$$

The interchange of summation and integral can readily be justified. Thus the transform of p is obtained through the Laplace transform of H. The embedded Markov chain of arrivals is determined through p but as the reader can see from Figure 7.9 the empty state is a special case. A customer departing and leaving the queue empty marks the start of an idle period.

The simplest way to analyse this chain is to work with the random variables directly. (Alternatively transform the equilibrium equations of the embedded Markov chain). Let Q_n denote queue length immediately following the nth departure, A_n the number of arrivals during the nth service then Equation 7.30 can be obtained, where I_n is 1 to allow for the departure of customer n-1 except if customer $n-1$ departs before customer n arrives, in which case I_n is 0, because $Q_{n-1} = 0$.

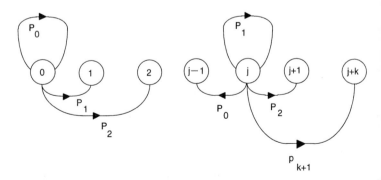

Figure 7.9 The embedded Markov chain of the M/G/1 queue

$$Q_{n+1} = Q_n + A_{n+1} - I_{n+1} \tag{7.30}$$

The probability generating function for Q_{n+1} is given by Equation 7.31.

$$Q_{n+1}(z) = E[z^{Q_{n+1}}] = E[z^{Q_n - I_{n+1}}] E[z^{A_{n+1}}]$$

$$= \left(\frac{E[z^{Q_n}] - q_{0,n}}{z} + q_{0,n} \right) H^*(\upsilon - \upsilon z)$$

$$= \left(\frac{Q_n(z) - q_{0,n}}{z} + q_{0,n} \right) H^*(\upsilon - \upsilon z) \tag{7.31}$$

Note that the number of arrivals during each service are independent of the current queue length, allowing this distribution to be factored out at the start of (Equation 7.31). By taking limits as $n \to \infty$, an equation for the equilibrium generating function, Q, for the number of customers left behind by one departing may be found from Equation 7.32.

$$Q(z) = \left(\frac{Q(z) - q_0}{z} + q_0 \right) H^*(\upsilon - \upsilon z)$$

$$= \frac{H^*(\upsilon - \upsilon z)(z - 1) q_0}{z - H^*(\upsilon - \upsilon z)} \tag{7.32}$$

Q exists whenever the queue is stable. However $-H^{*\prime}(0) = \frac{1}{\mu}$ is the mean service duration and so this condition is $\rho = -\upsilon H^{*\prime}(0) < 1$.

Setting $z = 0$ shows that q_0 is the equilibrium probability that a customer departs leaving the queue empty. Dividing by $z - 1$, taking limits as $z \to 1$ and applying L'Hopital's rule determines q_0 as in Equation 7.33 and hence Equation 7.34.

$$1 = Q(1) = \frac{q_0}{\lim_{z \to 1} \frac{z - H^*(\upsilon - \upsilon z)}{z - 1}} = \frac{q_0}{1 + \upsilon H^{*'}(0)} \quad (7.33)$$

$$q_0 = 1 - \rho, \, Q(z) = \frac{(1-\rho)H^*(\upsilon - \upsilon z)(z-1)}{z - H^*(\upsilon - \upsilon z)} \quad (7.34)$$

$Q(z)$, as derived, is the generating function for the number of customers left behind in queue by a departing customer. However this is then the equilibrium distribution for the number of customers which the customer finds on arrival.

This follows from the equality of equilibrium customer arrival distributions and customer departure distributions result described earlier.

Note also that Q gives the time equilibrium of customers in queue as well because of Poisson arrivals.

Q can be used to obtain the equilibrium waiting time transform. Let S be the total time a customer spends in the system (queuing time plus service time) with equilibrium Laplace transform $T^*(s)$.

By making an analogous argument to the one that we made for the distribution of the number of customers which arrive during a service, $Q(z)$ may be found.

This is just the number of arrivals during the customers' system time, so that Equation 7.35 may be obtained.

$$Q(z) = T^*(\upsilon - \upsilon z) \quad (7.35)$$

However the service time and waiting time are independent, giving Equation 7.36.

$$W^*(s) H^*(s) = T^*(s) \quad (7.36)$$

Putting these two together, and replacing $\upsilon - \upsilon z$ with s gives Equation 7.37 and Equation 7.38.

$$W^*(s) H^*(s) = T^*(s)$$

$$= \frac{(1-\rho) H^*(s)\left[\left(1-\frac{s}{\upsilon}\right)-1\right]}{\left(1-\frac{s}{\upsilon}\right)-H^*(s)} \tag{7.37}$$

$$W^*(s) = \frac{(1-\rho)s}{\upsilon H^*(s)+s-\upsilon} \tag{7.38}$$

This is the equilibrium transform for the waiting time.

7.3.4 The G/M/m queue in equilibrium

The analysis is very similar to the above but instead work with the queue immediately after each arrival. The waiting distribution follows immediately from this because the remaining time to complete the service of the customer at the front of the queue is exponential. (See the 'memoryless' property of the exponential above.) Thus if $r(z)$ is the generating function of customers in queue ahead of an arrival, the transform of the waiting distribution is then given by Equation 7.39 and Equation 7.40.

$$W^*(s) = r\left(\frac{\mu}{\mu+s}\right) \tag{7.39}$$

$$r(z) = \sum_{0}^{\infty} R_m z^m \tag{7.40}$$

The form of r is particularly simple for the G/M/m queue. Work the opposite way around from the M/G/1 queue and look at the number of departures between two arrivals. Suppose the interarrival density is given by $f(x)$. Then assume that there are $n > j$ customers in the queue. The probability that there are j departures before the next arrival is given by Equation 7.41 if $n - j \geq m$ so that all servers are busy during this period.

$$q_{n,j} = \int_0^\infty e^{-m\mu x} \frac{(m\mu x)^j}{j!} f(x) \, dx \tag{7.41}$$

For the $R_n, n \geq m - 1$, the equilibrium probability that an arrival finds n customers ahead of him in the queue is given by Equation 7.42, where the equation includes all the possibilities: the preceding customer may find $n - 1, n, n + 1, \ldots$ customers in the queue and then that there are $0, 1, 2, \ldots$ corresponding departures.

$$R_n = \sum_{k=0}^\infty R_{k+n-1} \int_0^\infty e^{-m\mu x} \frac{(m\mu x)^k}{k!} f(x) \, dx, \, n \geq m \tag{7.42}$$

These equations have the solution given by Equation 7.43 provided Equation 7.44 holds.

$$R_n = C \eta^n, \, n \geq m - 1 \tag{7.43}$$

$$\eta = F^*(m\mu - m\mu\eta) < 1 \tag{7.44}$$

Thus η is determined through the Laplace transform of F. A unique root exists if $m \mu \alpha < 1$ where α is the arrival rate. By setting $n = m-1$ we obtain Equation 7.45. In the special case where $m = 1$ the normalisation of Equation 7.46 gives Equation 7.47 and the queue length distribution is geometric.

$$C = R_{m-1} \eta^{1-m} \tag{7.45}$$

$$\sum_0^\infty R_n = 1 \tag{7.46}$$

$$R_0 = 1 - \eta \tag{7.47}$$

To obtain the remaining probabilities, in the case when $m > 1$, $R_{m-1}, R_{m-2}, ..., R_0$, compute the probability q_{nj} of there being j departures between two customer arrivals when there are n customers initially. This has been done already in the case $n - j \geq m$. Suppose $n \leq m$, then Equation 7.48 may be obtained, and if $n > m \geq n - j$, then Equation 7.49 is obtained.

$$q_{n,j} = \int_0^\infty \binom{n}{j} (1 - e^{-\mu x})^j e^{-\mu(n-j)x} f(x)\, dx \tag{7.48}$$

$$q_{n,j} = \int_0^\infty \binom{m}{n-j} e^{-\mu(n-j)x} \int_0^x \frac{m\mu (m\mu t)^{n-m-1}}{(n-m-1)!}$$

$$\times\ (e^{-\mu t} - e^{-\mu x})^{m-n+j}\, dt\, f(x)\, dx \tag{7.49}$$

The equilibrium probabilities that there are k customers found in queue on arrival for $k = m - 1, m - 2, \ldots$ are given by a series of equations such as Equation 7.50, Equation 7.51, and so on.

$$\begin{aligned}
R_{m-1} &= \sum_{j=0}^\infty q_{m+j-1,j}\, R_{m+j-2} \\
&= q_{m-1,0}\, R_{m-2} + R_{m-1} \sum_{j=1}^\infty \eta^{j-1}\, q_{m+j-1,j}
\end{aligned} \tag{7.50}$$

$$\begin{aligned}
R_{m-2} &= \sum_{j=0}^\infty q_{m+j-2,j}\, R_{m+j-3} \\
&= q_{m-2,0}\, R_{m-3} + q_{m-1,1}\, R_{m-2} + R_{m-1} \sum_{j=2}^\infty \eta^{j-2}\, q_{m+j-2,j}
\end{aligned} \tag{7.51}$$

These equations can be solved recursively by setting $R_{m-1} = 1$ and then determining $R_{m-2}, R_{m-3}, \ldots$ in turn. The solution is completed by renormalising the probabilities to sum to 1, as in Equation 7.52.

$$1 = \sum_{0}^{m-1} R_j + R_{m-1} \sum_{j=1}^{\infty} \eta^j = \sum_{0}^{m-1} R_j + R_{m-1} \frac{\eta}{(1-\eta)} \quad (7.52)$$

7.3.5 Results from the transforms

By differentiating Equation 7.38 and setting s to 0, the mean waiting time in a M/G/1 queue is obtained. This formula is known as the Pollazcek-Kinchine formula (Equation 7.53).

$$E[W] = \frac{\upsilon h_2}{2(1-\rho)} \quad (7.53)$$

Differentiating again the variance of the equilibrium wait may be determined, as in Equation 7.54.

$$\text{Var}[W] = \{E[W]\}^2 + \frac{\upsilon h_3}{3(1-\rho)} \quad (7.54)$$

In fact by a manipulation the waiting time transform can be inverted, as in Equation 7.55.

$$W^*(s) = \frac{(1-\rho)}{1 - \rho \frac{(1 - H^*(s))}{s/\mu}}$$

$$= (1-\rho) \sum_{k=0}^{\infty} \left[\rho \frac{(1 - H^*(s))}{s/\mu} \right]^k \quad (7.55)$$

However the Laplace transform of the residual service time density $r(x) = \mu(1 - H(x))$ is given by expression 7.56, so that Equation 7.57

gives the equilibrium waiting time density (Benes, 1956) in an M/G/1 queue.

$$\frac{(1 - H^*(s))}{s/\mu} \quad (7.56)$$

$$w(x) = (1 - \rho) \sum_{k=0}^{\infty} \rho^k r^{*k}(x) \quad (7.57)$$

For the G/M/1 queue the equilibrium distribution of queue length on arrival (and therefore on departure as well) is geometric, as in Equation 7.58.

$$\Pr\{Q = k\} = (1 - \eta)\eta^k \quad (7.58)$$

From this it follows immediately that the equilibrium G/M/1 waiting time distribution is exponential, as in Equation 7.59.

$$\Pr\{W \leq x\} = 1 - \eta\, e^{-\mu(1-\eta)x}, \quad x \geq 0 \quad (7.59)$$

The probability of not waiting at all is given by setting x to be 0, $1 - \eta$. This will not be the probability that the server is idle in general with an exception for the case when the arrival process is Poisson. In fact the waiting time distribution of the G/M/m queue, is always exponential with a point probability of not having to wait at all as the reader may verify.

7.4 Queuing networks

7.4.1 With general customer routes

The results of this section can be applied to both open and closed queuing networks. In an open queuing network customers make a sequence of visits to various nodes within the network before departing from the network altogether. In closed networks on the other hand the customer population within the network remains fixed with no new customers entering the network nor ones within in it departing. Examples of these kinds of network with yet more general servers can be found in Kelly (1979).

The results are similar in both cases but begin by considering the open case. The network is supposed to have J nodes. The server, at each node j, applies service at a rate which depends on the total number of customers present $\varphi_j(n_j)$, taken to be zero if there are no customers present. Each customer within the queue receives a part of this effort according to the service allocation vector $\gamma(k, n_j), k \leq n_j$. The amounts of service a customer requires at each queue are independent of one another and distributed exponential with unit mean. Some examples of how this vector might be set, along with the service rate, to obtain various kinds of server, are as follows:

1. ./M/1 queue with first come first served scheduling as in Equation 7.60.
2. ./M/m queue with first come first served scheduling as in Equation 7.61 and Equation 7.62.
3. Processor sharing as in Equation 7.63.

$$\varphi(n) = \varphi, \gamma(1, n) = 1, \gamma(j, n) = 0 \text{ otherwise} \tag{7.60}$$

$$\varphi(n) = n \varphi \, \gamma(j, n) = \frac{1}{n}, j \leq n \leq m \tag{7.61}$$

$$\varphi(n) = m \varphi \; n > m,$$

$$\gamma(j, n) = \frac{1}{m}, j \leq m, \gamma(j, n) = 0 \text{ otherwise} \tag{7.62}$$

$$\varphi(n) = \varphi, \; n \geq 1, \gamma(j, n) = \frac{1}{n}, j \leq n \tag{7.63}$$

The customers' routes through the network are determined by an ordered set of nodes, which may be regarded as the list of nodes

Queuing theory

which the customer will visit in the order in which he will visit them. A list is denoted $r = (r(1), .., r(S(r)))$ where $S(r)$ is the number of nodes visited. Repeat visits are permitted.

It is supposed that customers following route r arrive as a Poisson stream with rate υ_r, independently. Furthermore it will be supposed that the number of routes is finite although this is not a necessary assumption.

Apart from specifying which order the customers of a particular routing class visit a queue it is also necessary to specify how they enter each queue. This is determined by the random entrance vector, δ, which is independent of routeing class. Given that there are n_j customers in the queue immediately prior to the entry of the new customer the position of the new customer is given by $\delta(k, n_j + 1)$. By choice of this vector it is possible to model queues with last come first served scheduling or service in random order.

It remains to specify the state of the network. It is not enough simply to stipulate the number of customers in each queue since this ignores which class of customers are in what positions. The state of node j is given by a vector $c_j(n_j)$ where the kth component of c_j is determined by the routeing class of the customer in position k together with the stage which he has reached along his route. The state of the network consists of a vector whose components are the individual queue state vectors, C. Let $q(C,D)$ be the transition rate from state C to state D.

There are three kinds of transition within the network, an arrival to the network, a customer moving between queues and a departure. The new state of the network immediately after the arrival of a customer, following route r, into position k of its initial queue is given by $A_k^r(C)$. The state of the network immediately after the departure of a customer on route r from position k at its last queue is $L_k^r(C)$. The state of the queue immediately after a customer following route r moves from queue i position e to queue h position k is $M^{r_{ie,hk}}(C)$. Denote by j the first queue on route r and g the last queue on that route, then these transitions take place at the rates given in Table 7.4.

150 Queuing networks

Table 7.4 Queuing network transition rates

Transition	Rate
Arrivals	$\upsilon_r \delta(k, n_j+1)$
Departures	$\varphi_g(n_g) \gamma(k, n_g)$
Moving between queues	$\gamma(e, n_i) \varphi_i(n_i) \delta(k, n_h+1)$

The equilibrium queue distribution is of product form, as in Equations 7.64 to 7.66, where G_j is a normalisation constant, a_j is the total traffic arriving at node j, and is given by Equation 7.67, and $\alpha_j(r)$ is the number of visits that customers on route r make to node j and (with a little abuse of notation) $\upsilon_{r(k)}$ is the arrival rate of the class of the customer who is currently in position k.

$$\pi(C) = \prod_j \pi_j(c_j) \tag{7.64}$$

$$\pi_j(c_j) = G_j \prod_{k=1}^{n_j} \frac{\upsilon_r(k)}{\varphi_j(k)} \tag{7.65}$$

$$G_j^{-1} = \sum_{n=0}^{\infty} \frac{a_j^n}{\prod_{k=1}^{n} \varphi_j(k)} \tag{7.66}$$

$$a_j = \sum_r \alpha_j(r) \upsilon_r \tag{7.67}$$

The following result makes it easy to verify that Equation 7.64 is indeed the equilibrium queue distribution.

Theorem 1

Suppose that numbers $q'(C,D) > 0$ for each pair of states C,D can be found so that Equation 7.68 is obtained.

$$\sum_{D \neq C} q'(C,D) = q'(C) = q(C) = \sum_{D \neq C} q(C,D) \qquad (7.68)$$

Furthermore, suppose numbers $\pi(C) > 0$ for each state C can be found satisfying Equation 7.69 for each pair of states C,D. Then π is the equilibrium probability distribution for both Markov process whose transition rates are determined by $q(C,D)$, $q'(C,D)$ respectively.

$$\pi(C) q(C,D) = \pi(D) q'(D,C) \qquad (7.69)$$

It is sufficient to obtain the equilibrium equations from Equation 7.68, as in Equation 7.70, but the left hand side is given by Equation 7.71.

$$\sum_{D \neq C} \pi(C) q'(C,D) = \sum_{D \neq C} \pi(D) q(D,C) \qquad (7.70)$$

$$\pi(C) q'(C) = \pi(C) q(C)$$

$$= \sum_{D \neq C} \pi(C) q(C,D) \qquad (7.71)$$

In order to make use of this result we must find suitable transition rates $q'(C,D)$. This is done by constructing the time reversed queuing network which has transition rates given by Table 7.5.

In the reversed network customers start from their last queue and follow their route backwards, and also the roles of γ and δ are swapped. Thus the first queue on route r is g, the last j and the queue transition is from queue h position k to queue i position e.

152 Queuing networks

Table 7.5 Transition rates for the reversed network

Transition	Rate
Arrivals	$\upsilon_r \gamma(k, n_g + 1)$
Departures	$\varphi_j(n_j) \delta(k, n_j)$
Moving between queues	$\gamma(1, n_i + 1) \varphi_j(n_h) \delta(k, n_h)$

Consider the probability flux of arrivals of customers on route r to their first node j, $r(1) = j$ it is easily verified that Equation 7.72 holds and, by symmetry, a similar relationship holds for departures, as in Equation 7.73.

$$\upsilon_r \delta(k, n_j + 1) \pi(C) \\ = \varphi_j(n_j + 1) \delta(k, n_j + 1) \pi(A_k^r(C)) \quad (7.72)$$

$$\varphi_g(n_g + 1) \gamma(k, n_g + 1) \pi(C) \\ = \upsilon_r \gamma(k, n_g + 1) \pi(L_k^r(C)) \quad (7.73)$$

Similarly for transitions between queues from node i position e to node h position k following route r, Equation 7.74 is also easily verified.

$$\varphi_i(n_i) \gamma(e, n_i) \delta(k, n_h + 1) \pi(C) \\ = \varphi_h(n_h + 1) \delta(k, n_h + 1) \gamma(e, n_i) \pi(M^{r_{ie,hk}}(C)) \quad (7.74)$$

By adding Equations 7.72 to 7.74 and summing over all possible transitions out of state C, Equation 7.68 may be obtained. The total rate out of state C in both networks is given by Equation 7.75.

$$q(C) = q'(C) = \sum_{j=1}^{J} \varphi_j(n_j) + \sum_r \upsilon_r \quad (7.75)$$

Queuing theory

Thus the conditions of theorem 1 are verified and Equation 7.64 is indeed the equilibrium distribution of the network.

As described so far the customers have to call at each node in turn along their route. However it is possible to model customer classes in which each new customer is assigned one from a possible set of routes at random. This is achieved as follows. Let s be a route which can be followed. Suppose the probability that this route is followed is f_s. Each such route s is then offered traffic $\upsilon_s = \upsilon f_s$ where υ is the total traffic offered by the customer class using these random routes.

The approach to closed queuing networks is very similar. To begin with customers may be regarded as following a route given by a list of nodes in the order in which they are to be visited. However when a customer departs the final node in its list it returns to the first and cycles through the network again. For each route there is a population of customers which follow it, N_r. Let N denote the population vector giving the number of customers following each route r.

The above definitions resemble those for an open network but with no departures or arrivals from the network. A reasonable candidate for the equilibrium is to use the same form as the open network. However there are no arrivals and so a set of dummy arrival rates are assigned to each route υ_r. The precise values are not important as long as they are positive. The reason for this freedom in choice of arrival rates is mentioned in a moment.

Only certain states of the corresponding open network can be attained in the closed network. Thus, to obtain the candidate closed equilibrium distribution, the open distribution is renormalised to the total probability of all the possible states T in the closed queuing network, as in Equation 7.76 where $\pi(C)$ is defined in Equation 7.64 so that the equilibrium probability of being in state C is given by expression 7.77.

$$G^{-1}(N) = \sum_{C \varepsilon T} \pi(C) \qquad (7.76)$$

$$G(N)\pi(C) \qquad (7.77)$$

154 Queuing networks

That this is indeed the equilibrium may be verified using the same argument based on theorem 1. Since the system is closed the equilibrium distribution cannot depend on the dummy arrival rates. The normalisation constant cancels out the dependence of the equilibrium distribution on the dummy arrival rates.

As an example, consider a simple model of a series of transmission links access to which is regulated by window flow control, see Figure 7.10. As depicted, a packet gains access by obtaining one of the window flow control tokens. This token is carried with the packet over the three transmission links before the token is released. It is supposed to return immediately to be available for other packets to use. As Figure 7.10 shows this can be represented by a closed cyclic queuing network with the tokens as customers. These are 'served' by the arrivals which take place at rate λ and then by the transmission links. Arrivals which find no tokens are lost.

Suppose that there are N tokens and that the service rate at each of the transmission links is φ. The network state is represented by Equation 7.78.

$$n = (n_1, n_2, n_3, n_4), n_1 + n_2 + n_3 + n_4 = N \qquad (7.78)$$

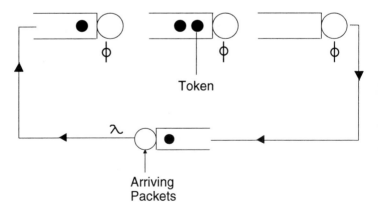

Figure 7.10 Window flow control over three consecutive transmission links

Using Equation 7.64, the equilibrium distribution is then given by Equation 7.79, where G is given by Equation 7.80.

$$\pi(n) = G^{-1} \frac{1}{\varphi^{n_1}} \frac{1}{\varphi^{n_2}} \frac{1}{\varphi^{n_3}} \frac{1}{\lambda^{n_4}} \tag{7.79}$$

$$G = \sum_{n:n_1+n_2+n_3+n_4=N} \frac{1}{\varphi^{n_1}} \frac{1}{\varphi^{n_2}} \frac{1}{\varphi^{n_3}} \frac{1}{\lambda^{n_4}} \tag{7.80}$$

The probability a packet is lost because there is no token available is given by expression 7.81.

$$G^{-1} \sum_{n:n_4=0} \frac{1}{\varphi^{n_1}} \frac{1}{\varphi^{n_2}} \frac{1}{\varphi^{n_3}} \tag{7.81}$$

As a second example consider the communication network shown in Figure 7.11. Customers move from A to C via the central queue or from B to D via the central queue with arrival rates υ and λ respectively. The service rate at each queue is φ.

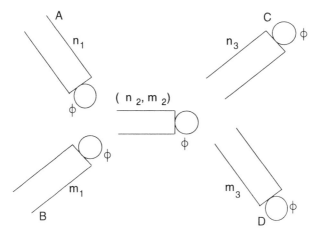

Figure 7.11 Communication network with two customer routes

Let n_i denote the number of customers going from A to D at each stage i and m_i the number at each stage going from B to C then, again using Equation 7.64, the equilibrium distribution is as in Equation 7.82.

$$\pi(n,m) = G^{-1} \left(\frac{\upsilon}{\varphi} \right)^{n_1} \left(\frac{\lambda}{\varphi} \right)^{m_1} \binom{n_2+m_2}{n_2}$$

$$\times \left(\frac{\upsilon}{\varphi} \right)^{n_2} \left(\frac{\lambda}{\varphi} \right)^{m_2} \left(\frac{\upsilon}{\varphi} \right)^{n_3} \left(\frac{\lambda}{\varphi} \right)^{m_3} \qquad (7.82)$$

Note the combinatorial term for the centre queue. The state reflects only the number of customers of each type at each queue and the combinatorial term accounts for all possible arrangements, which are all equally likely. (Equation 7.83.)

$$G^{-1} = \left(1 - \frac{\upsilon}{\varphi} \right)^2 \left(1 - \frac{\lambda}{\varphi} \right)^2 \left(1 - \frac{\upsilon}{\varphi} - \frac{\lambda}{\varphi} \right) \qquad (7.83)$$

7.4.2 Fixed point models for closed queuing networks

In many communication applications, the distributions of system random variables may be known only rather imperfectly. Consider the distribution of sojourn (queuing plus transmission) time of a packet at a node within a communication network. This varies with the flow of packets through the node. It is possible to have a very good estimate of how the mean sojourn time depends on this quantity but not have more detailed information about the distribution than this.

In these circumstances detailed modelling of the sojourn time distribution appears inappropriate. As will now be shown it is possible to deduce a great deal about performance without having to make further assumptions which would be difficult to justify and which may also complicate analysis (Kelly, 1989). These models are

Queuing theory

capable of considerable generalisation including having priority queues although this extension will not be discussed here.

The nodes within the network are labelled 1, 2,..., J. Each customer class, r, has N_r customers in it. For each such class there is a rule which determines how the customers in the class move around within the network. This rule is based on cycles with one being completed every time certain sequences of visits are executed.

A simple rule of this type is given if a route is determined by a list of nodes which have to be visited in order. A cycle is completed once all the nodes in the list have been visited. Note that repeat visits to nodes within the list are permitted.

A more general rule is given if it is supposed that routing in each class is determined by an irreducible homogeneous Markov chain. Let π_{jk} denote the probability that the customer visits node k immediately after departing node j. A cycle is completed when a given node is returned to.

As can be seen the routing rules include those mentioned in the previous section.

Given routing rules such as these, let the mean number of visits to each node per cycle be given by α_{jr}. Finally denote the mean rate at which class r complete cycles by θ_r and the flow of customers through node j by ρ_j. Conservation of flow gives ρ_j in terms of the θ_r's, as in Equation 7.84.

$$\rho_j = \sum_j \alpha_{jr} \theta_r \qquad (7.84)$$

The model is determined once the mean sojourn time at each node is specified. In the simplest cases this will be a function of mean flow and service capacity only. (Equation 7.85.)

$$D_j = D_j (\rho_j ; \varphi_j) \qquad (7.85)$$

This definition allows all the queuing models described in Sections 7.2.3 and 7.3.3 to 7.3.5.

The conservation of flow equation (Equation 7.85), together with Little's law applied to each customer class, gives Equation 7.86, which yield $J + R$ equations in $J + R$ unknowns.

$$N_r = \theta_r \sum_r \alpha_{jr} D_j(\rho_j; \varphi_j) \tag{7.86}$$

If D_j is non-negative and an increasing function of ρ_j then the solution is unique. We now demonstrate this.

Define H as in Equation 7.87.

$$H = -\sum_j \int_0^{\rho_j} D_j \, d\rho_j' + \sum_r N_r \log \theta_r \tag{7.87}$$

However H is strictly concave because the D_j's are continuous and monotonic increasing. Furthermore any maximum of H must be unique since H is strictly concave. In addition the maximum must be located at an interior point. This is therefore determined by the zeros of the derivative and must be satisfied at only that point.

H has a derivative as in Equation 7.88.

$$\frac{\partial H}{\partial \theta_r} = -\sum_j \alpha_{jr} D_j + \frac{N_r}{\theta_r} \tag{7.88}$$

It can now be seen that Equation 7.86 must have just one solution since the solution determines the unique maximum of H.

Repeated substitution is one of the most numerically efficient ways of solving the above sets of equations. One begins with an initial value of the ρ_j's, for example as in Equation 7.89. The θ_r's and ρ_j's may then be solved recursively as in Equations 7.90 and 7.91.

$$\rho_j^{(1)} = 0, \, j = 1, \ldots, J \tag{7.89}$$

$$\theta_r^{(n)} = \frac{N_r}{\sum_r \alpha_{jr} D_j(\rho_j^{(n)}; \varphi_j)} \quad (7.90)$$

$$\rho_j^{(n+1)} = \sum_j \alpha_{jr} \theta_r^{(n)} \quad (7.91)$$

On many occasions the above recursion will not converge. However the use of damping gets over this problem. With damping factor γ, $0 < \gamma < 1$ Equation 7.90 can be written as in Equation 7.92.

$$\theta_r^{(n+1)} = \frac{\gamma N_r}{\sum_r \alpha_{jr} D_j(\rho_j^{(n)}; \varphi_j)} + (1-\gamma) \theta_r^{(n)} \quad (7.92)$$

On some occasions heavy damping may be needed i.e. values of γ close to 0.

As an example, consider again the model of a series of transmission links, access to which is regulated by window flow control, see Figure 7.10. Again there are N tokens which circulate at rate θ. From Figure 7.10 this is determined by the fixed point Equation 7.93.

$$N = \frac{\theta}{(\lambda - \theta)} + \frac{3\theta}{(\varphi - \theta)} \quad (7.93)$$

The rate of transmitting packets is θ of course and therefore the proportion of packets which are lost is given by expression 7.94.

$$1 - \frac{\theta}{\lambda} \quad (7.94)$$

7.5 Multi-access channels

Queuing analysis has been successfully applied to determine the throughput of a wide range of access protocols used in radio communications and elsewhere. The following provide two examples illustrating a typical approach to the problem of determining the throughput.

7.5.1 ALOHA models

Consider first asynchronous (unslotted) ALOHA, as shown in Figure 7.12. All packets are the same length and the transmission of a packet will be successful provided another packet does not transmit in it's vulnerable period. The probability of a successful transmission is therefore given by Equation 7.95, where γ is the total transmission attempt rate.

$$\Pr\{\text{Successful transmission}\}$$
$$= \Pr\{\text{No transmission in vulnerable period}\} = e^{-2\gamma} \quad (7.95)$$

The total throughput is the rate of making successful transmissions, as in Equation 7.96.

Throughput = Transmission attempt rate
 × Probability of success

$$= \gamma e^{-2\gamma} \quad (7.96)$$

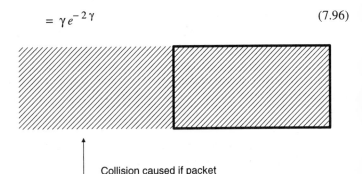

Collision caused if packet transmits here

Figure 7.12 Vulnerable period for unslotted ALOHA

This is maximum for $\gamma = 1/2$ so that the throughput cannot exceed $1/2e$. Note that the total rate of transmissions and retransmissions is being modelled as a Poisson process of rate γ. Even if the process of fresh transmissions is Poisson, the assumption that the retransmissions are independent Poisson can only be justified if the retransmissions take place at random over long intervals following the collisions which give rise to them.

The throughput can be increased by adopting a slotted version of this protocol so that packets can only be transmitted at the beginning of each slot which was depicted in Figure 7.2. In this way if a collision takes place the two packets coincide completely, eliminating the possibility of packet loss by partly overlapping packets. The probability of success becomes $e^{-\gamma}$ and the throughput $\gamma e^{-\gamma}$. The maximum throughput becomes $\frac{1}{e}$ at $\gamma = 1$.

The above results must be treated with caution. The conclusions are correct but only if they are qualified. The reason for this is that the analysis does not take into account the nature of the underlying Markov chain which it purports to analyse. Indeed if an infinite station model is adopted the equilibrium can be shown not to exist (Kelly, 1985)! The slotted ALOHA channel will jam with probability 1.

Even in the case where the station population is finite the channel may perform inadequately. This is because the system may have bistable operation with one stable point at low traffic and most stations idle giving high throughput and a second one where most stations are busy and the throughput is low because of the large number of stations all attempting together. The system may remain in this second state for long periods.

The underlying difficulty is that the protocol does not adapt the number of transmission attempts according to the number of busy stations. The retransmission probability should be high if the number of busy stations is small but low when there are large numbers of busy stations. This problem is overcome by having the stations keep an estimate of the backlog which is updated following each slot according to whether it was idle contained a successful transmission or contained a collision (Gallager and Bertsekas, 1987).

7.5.2 Non-Persistent Carrier Sense Multiple Access

Under Carrier Sense Multiple Access (CSMA) mobiles in a communication network having packets to transmit do so if they 'sense' the radio channel is not being used. We do not go into a description of how this is achieved but suppose that all mobiles become aware of a packet transmission within τ time units after it was begun. The period τ is called the vulnerable period.

The protocol has only two steps:

1. The channel is 'sensed'. If it is found to be idle the packet is transmitted immediately.
2. If the channel is 'sensed' busy the packet transmission is rescheduled using the retransmission distribution. At the end of the delay period the protocol moves to step 1.

Figures 7.13(a) and 7.13(b) show successful and unsuccessful transmissions respectively. The analysis for non-persistent CSMA is as follows. During the start of a transmission there is a short period during which other transmissions may start before all other mobiles become aware that the transmission has started. This interval, in units of packet transmission time is τ as already mentioned. Making the usual assumption that transmission attempts are taking place at rate λ as a Poisson process, the probability that the packet is transmitted successfully is that there are no further transmission attempts during this vulnerable period, which is given by expression 7.97.

$$e^{-\tau\lambda} \tag{7.97}$$

Contending packets may arrive at any time during the initial vulnerable period and it is the last of these arrivals which determines when the contention period ends and the line goes idle. Let Z be a random variable defined as in Equation 7.98.

$$Z = \begin{cases} 0 & \textit{No collisions} \\ \textit{Time between first and last colliding packet} \end{cases} \tag{7.98}$$

Queuing theory 163

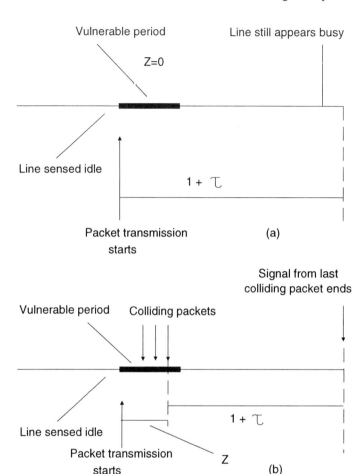

Figure 7.13 Transmission under the non-persistent CSMA protocol: (a) successful; (b) unsuccessful

Observe that Z must be less than τ. Given that there is at least one colliding packet the distribution of time between the last packet to arrive and the end of the vulnerable period is the same as the time until the first packet transmission. This observation shows that the distribution of Z is given by expression 7.99.

164 Multi-access channels

$$e^{-\lambda(\tau-z)} \tag{7.99}$$

From Equation 7.99 the mean value of Z is obtained, as in Equation 7.100.

$$E[Z] = \tau - \frac{1}{\lambda} + \frac{e^{-\lambda\tau}}{\lambda} \tag{7.100}$$

Packet transmissions and contention go in cycles. There are only two kinds:

1. Successful transmission, equal to the idle period plus packet transmission time.
2. Unsuccessful transmission, equal to the idle period plus the contention period.

The mean duration of an idle period is $\frac{1}{\lambda}$.

The definition of Z takes into account both contention and a successful transmission. The mean for this second part of the cycle is given by expression 7.101.

$$1 + \tau + E[Z] \tag{7.101}$$

The system throughput C is then determined as in Equation 7.102.

$$C = \frac{e^{-\tau\lambda}}{1 + \tau + E[Z] + \frac{1}{\lambda}} = \frac{\lambda e^{-\tau\lambda}}{\lambda + 2\lambda\tau + e^{-\tau\lambda}} \tag{7.102}$$

The maximum possible throughput can now be found from examining C over all possible values of λ.

Other forms of CSMA protocol are discussed in Kleinrock (1975b).

Note, once again, that these results must be treated cautiously; stability questions arise with CSMA protocols as well.

7.6 References

Benes V.E. (1956) On Queues with Poisson Arrivals, *Annals of Mathematical Statistics*, **28**, pp. 670 677.

Feller W. (1958) *An Introduction to Probability Theory and Its Applications*, **1**, 3rd edn., Wiley.

Feller W. (1970) *An Introduction to Probability Theory and Its Applications*, **2**, Wiley.

Gallager R.G. (1977) A Minimum Delay Distributed Routing Algorithm, *IEEE Transactions on Communications*, **Com 23** (1) January, pp. 73-85.

Gallager R.G. and Bertsekas D.M. (1987) *Data Networks*, Prentice-Hall.

Kelly F.P. (1979) *Reversibility and Stochastic Networks*, Wiley.

Kelly F.P. (1985) Stochastic Models of Computer Communication Systems, *Journal Royal Statistical Society*, Series B, **47** (3), pp. 379 -395.

Kelly F.P. (1989) On a Class of Queuing Approximations for Closed Queuing Networks, *Queuing Systems*, **4**, pp. 69-76.

Kleinrock L. (1975a) *Queuing Theory*, **1**, Wiley Interscience.

Kleinrock L. (1975b) *Queuing Theory*, **2**, Wiley Interscience.

Tackacs L. (1962) *An Introduction to the Theory of Queues*, Oxford University Press.

8. Information theory

8.1 Introduction

In this chapter, the ideas associated with information and entropy are considered and related to probability defined in terms of incomplete knowledge. Mutual and self information are introduced and these concepts are applied to various models for communications channels.

The concepts of Information and the related quantity Entropy, date back to the early days of thermodynamics. The concept of entropy arose in considerations connected with the theory of heat, and its relationship to information was not at first realised. However, in retrospect it is not surprising that the analytical treatment of information should have its origin in this area of physical science, for the subject of thermodynamics is primarily concerned with the determination of the laws that govern the conversion of mechanical and other forms of energy, into heat energy. Heat is characterised by its disorder, and ultimately by the irreversibility of certain processes which involve heat transfer. The degree of organisation of a system can also be interpreted as a measure of the quantity of information incorporated in it. This idea, was developed by Szilard in 1929, after which time the notion of entropy was another term for information, and that the physical measure of entropy was also a measure of the quantity of information or degree of organisation of the corresponding physical system. More precisely, the difference in the quantity of information between two states of a physical system is equal to the negative of the corresponding difference in entropy of the two states.

In nature it is found that self contained systems change from more highly organised structures to less organised ones, i.e. from states of higher information content to states of lower information content.

It is important to note that in the above context a physical system corresponds to many different thermodynamic systems, and entropy is not simply a property of the system, but of the experiments chosen to undertake on the system. For example, one normally controls a set of variables for the physical system, and one measures the entropy for that set. A solid with N atoms has approximately $6N$ degrees of freedom, of which only a few, e.g. pressure, temperature, magnetic field, are usually specified to obtain the entropy. By increasing this set, one obtains a sequence of entropy values, each corresponding to the chosen constraints.

According to Jaynes (1978), the entropy of a thermodynamic system is a measure of the degree of ignorance of a person whose sole knowledge about its microstate consists of the values of the macroscopic quantities, e.g. pressure and temperature, which define its thermodynamic state, and it is a completely objective quantity in the sense that it can be measured in the laboratory.

A very good source of reference concerning both the thermodynamic and the information processing approach to entropy and information can be found in the recent book edited by Leff and Rex (1990).

Before we concentrate on information theory proper, it will be worth laying a few foundations concerning the role of probability theory in information, and in trying to use probability theory in information theory, one immediately sees that there exist many different views as to what probability theory is, a sample of the various views being:

1. Kolmogorov; The theory of additive measures.
2. Jeffreys; The theory of rational belief.
3. Fisher; The theory of frequencies in random experiments.
4. de Morgan; The calculus of inductive reasoning.
5. Laplace; Common sense reduced to calculation.
6. Bernoulli; The art of conjecture.
7. von Mises; The exact science of mass phenomena and repetitive events.

For over one hundred years, controversy has raged between those holding these, and other views. However, these views just reflect the particular problems that were being addressed by the authors, and one should take the view that the above are valid and useful in different contexts, but that some views are more general than others, and it is this general approach that will now be discussed before we apply it to information theory.

According to the view put forward by Jeffreys, probability expresses a state of knowledge about any system under investigation, and as such applies to a very wide range of problems. This approach to probability theory incorporates many of the above mentioned interpretations.

It should not be considered wrong to adopt a narrow view of probabilities in terms of random experiments and frequencies of occurrence, but there is certainly nothing to be lost in adopting a broader view in terms of states of knowledge, and if the problem requires the concept of relative frequencies etc., this should emerge out of the broader formulation.

Consider propositions EA and EB. Using Boolean algebra, we may construct new propositions from A and B by conjunction, disjunction and negation, as in Equations 8.1 to 8.3.

$$A B = \text{Both } A \text{ and } B \text{ are true} \tag{8.1}$$

$$A + B = \text{At least one of the propositions is true} \tag{8.2}$$

$$\tilde{A} = A \text{ is false} \tag{8.3}$$

The conditional probability that A is true given that B is true, is a real number between 0 and 1, and is represented by the symbol $P(A|B)$. Therefore $p(A+B|CD)$ is the probability that at least one of the propositions (or hypotheses) A,B, is true, given that both C and D are true, etc.

In this formulation of probability as a state of knowledge, or a representation of incomplete knowledge, all probabilities are conditional on some information, and there does not exist an absolute probability.

The rules for conducting scientific inference follow from the simple laws of probability as in Equations 8.4 and 8.5.

$$\begin{aligned} P(AB|I) &= p(A|I)p(B|AI) \\ &= p(B|I)p(A|BI) \end{aligned} \tag{8.4}$$

$$p(A|B) + p(\tilde{A}|B) = 1 \tag{8.5}$$

If $p(B|I) \neq 0$ then Bayes' theorem follows as in Equation 8.6, where I stands for any prior information.

$$p(A|BI) = \frac{(pA|I)p(B|AI)}{p(B|I)} \tag{8.6}$$

Using Bayes' theorem, it is possible to incorporate chains of evidence into one's reasoning, the posterior probability becoming the prior for the next iteration as new data become available. This method of inference goes back to Laplace in the 18th century, but the mathematics necessary to show that Bayesian inference is the only logical way to proceed was given in 1946 by R.T. Cox. However, the formalism does not show us how to assign the initial probabilities. There are several methods that may be used to assist in the initial assignment of probabilities, the first one to consider being symmetry. For example, if one considers coins or dice that are unbiased, it would seem reasonable to express this state of knowledge by assigning equal probabilities to the allowed states, and this allows us to make the best predictions we can from our state of incomplete knowledge. This is referred to as Laplace's law of insufficient reason.

Boltzmann in 1877 wished to determine how gas molecules distribute themselves in a conservative field such as gravity. He divided the 6-d 'phase space', 3 spatial dimensions and 3 momentum dimensions, into equal cells, with N, molecules present in the ith cell.

The number of ways this distribution can be realised is as in Equation 8.7.

$$W = \frac{N!}{N_1! N_2! \ldots N_n!} \tag{8.7}$$

The prior knowledge that Boltzmann was able to use was that the total energy and the total number of molecules in the gas was constant, as in Equations 8.8 and 8.9.

$$N = \sum_i N_i = \text{constant} \qquad (8.8)$$

$$E = \sum_i N_i E_i = \text{constant} \qquad (8.9)$$

Boltzmann considered that the most probable distribution for this case was the one that maximises W subject to the constraints given above.

If the total number of gas molecules is large, then Stirling's approximation gives Equation 8.10, which may be seen to be related to the Shannon entropy.

$$\frac{1}{N} \log W = -\sum_i \left(\frac{N_i}{N}\right) \log\left(\frac{N_i}{N}\right) \qquad (8.10)$$

Thus the most likely distribution is obtained by maximising entropy subject to the constraints of the problem. This methodology has, over the years, been applied to a vast range of problems, with remarkable success.

Information theory says that a random variable, X, which has an associated probability density function, $p(x)$, has an entropy given by Equation 8.11 which means that H bits are sufficient to describe X on the average.

$$H = -\sum p(x) \log p(x) \qquad (8.11)$$

Kolmogorov similarly described algorithmic complexity, $K(x)$, to be the shortest binary program that describes X. According to Kolmogorov, Information theory must precede probability theory, and should not be based on it, and his contribution to information theory followed by a direct development of his ideas in algorithmic complexity. In particular, he was interested in finding any determinism in random events, and his work on turbulence must be seen in this light as an attempt to find deterministic order in chaotic processes, and

Information theory 171

indeed turbulence was one of the key phenomena that motivated the resurgence of interest in non-linear dynamical systems, and the range of chaotic phenomena and strange attractors.

It is interesting that many of the ideas borrowed from information theory are used to classify many of the chaotic signals, such as heart beats, brain waves, chemical reactions, lasers, flames, etc., and Dimensions, Entropies and Lyapunov exponents are used routinely.

8.2 Information capacity of a store

Figure 8.1(a) shows a system with two possible states. If the position of the points is unknown, a priori, and we learn that the point is in the left hand box, say, we gain information amounting to 1 bit. If we obtain this information, we save one question in order to locate the point. Hence the maximum information content of a system with two states is one bit.

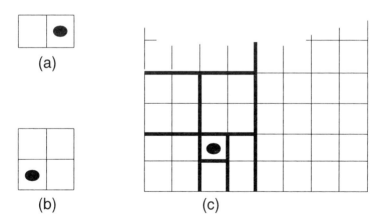

Figure 8.1 Information capacity of a store: (a) a box with two states; (b) it takes two questions and their answers to locate a point in a system with four states — right or left, up or down? (c) in order to locate a point on a board with $64 = 2^6$ states, six questions are required

For a box with two states, one needs two questions in order to locate the point, (such as, is the point to the right or left and up or down?) and the maximum amount of information is two bits. This can, of course, be written as the logarithm to the base 2 of the number of allowable states. In general, the maximum information content associated with a system having N states, is given by Equation 8.12.

$$I = \log_2 N \tag{8.12}$$

We can also consider the outcome of statistical events, and calculate the average gain of information associated with such events. As an example, consider coin tosses, such that the outcome, heads or tails, occurs with equal probability of 1/2.

The information gain acquired by learning that the outcome of an experiment is heads, say, is given by Equation 8.13.

$$I = -\left(\frac{1}{2}\log_2 \frac{1}{2} + \frac{1}{2}\log_2 \frac{1}{2}\right) = 1 \tag{8.13}$$

In general, therefore, the information gain associated with many states having probabilities associated with each state, is given by Equation 8.14.

$$I = -\sum_i p_i \log_2 p_i \tag{8.14}$$

8.3 Information and thermodynamics

Boltzmann realised that entropy was a measure of disorder of the configuration of the system being considered. This viewpoint was extended by Szilard in 1929, and he identified entropy with information, and the measure of information with the negative of the measure of entropy. This laid the foundations of information theory.

This viewpoint came about due to a paradox conceived of by Maxwell concerning two chambers separated by a common partition which could be removed to allow objects (gas molecules) to move freely between the two chambers. Normal experience tells us that if one of the chambers initially has gas present and the partition is withdrawn, then gas will rapidly diffuse to fill both chambers. The reverse never occurs.

Maxwell proposed his famous 'demon' whose purpose was to open a small frictionless door present in the partition, and when a particle, due to thermal motion heads towards the door, the demon opens the door to let the particle through the partition, in one direction only, and as time passes, one chamber gets filled up with particles at the expense of the other, thus seeming to violate the second law of thermodynamics. This paradox was explained by Szilard, who pointed out that in order for the demon to perform, it had to be very well informed about the position and velocity of the particle as it approached the door. Szilard's argument is very well given in the book by Leff and Rex, (1990), and led to the realisation of the connection between entropy and information.

8.3.1 Entropy of finite schemes

Consider a complete system of events, $A_1, A_2, \ldots A_n$ such that one and only one event may happen at a given trial, e.g. tossing a coin, for which $n = 2$, and the events are mutually exclusive.

Every finite scheme describes a state of uncertainty, and we only know the probability of possible outcomes, i.e. a one in six chance of a particular face of a dice being uppermost (for a fair dice).

If for one scheme the particular probabilities are different, then there will be different degrees of uncertainty associated with states having different probabilities. It is therefore essential to derive a measure of uncertainty of a particular scheme, and the quantity given in Equation 8.15 is a very suitable measure.

$$H(p_1, p_2, \ldots p_n) = -\sum_{i=1}^{n} p_i \log p_i \tag{8.15}$$

The logarithm is taken to be an arbitrary base, and $p_i \log p_i = 0$ if $p_i = 0$. The quantity H is called the entropy of the finite scheme.

Around the time when Szilard was considering Maxwell's demon and the connection between entropy and information, the whole of physics was undergoing vast changes due to the development of quantum theory.

Of particular interest is the work of J. von Neumann, who in 1932, showed that in the process of making measurements, the observer must be taken into account, and this gathered together ideas of entropy, information and irreversibility.

It was R.V. Hartley (1928), of the Bell Telephone laboratories who gave the logarithmic definition of information from a communications viewpoint, and it was C. Shannon who further refined the ideas in 1949.

Suppose that X is a random variable or random vector that can be described by its probability density function $p(x)$, where x takes values in a certain alphabet, A_x. Shannon (1949), defined the entropy of X by Equation 8.16.

$$H = -\sum_{x \varepsilon A_x} p_X(x) \log p_X(x) \tag{8.16}$$

Similarly, given two discrete random variables X and Y, one can define various entropies $H(X,Y)$, $H(X)$, $H(Y)$ as well as the conditional entropy as in Equation 8.17 (see the next section).

$$H(X|Y) = H(X,Y) - H(Y) \tag{8.17}$$

It should also be mentioned that the statistician R.A. Fisher in 1922 defined the information content of a statistical distribution, and this Fisher information still plays a fundamental role in statistical estimation and parameter estimation in general, and allows one to put bounds on the accuracy with which one can estimate parameters.

8.4 Mutual and self information

Let X and Y be two discrete variables with possible outcomes x_i, $i = 1,2,...,n$, and y_i, $i = 1,2,...,m$, respectively. If an outcome $Y = y_j$

is observed, and we wish to quantitatively determine the amount of information that this observation provides about the event $X = x_i$, $i = 1,2,...n$, then the appropriate measure of information has to be selected. If X and Y are statistically independent, then the occurrence of $Y = y_j$ provides no information about the occurrence of $X = x_i$.

If, however, the random variables are fully dependent on one another, then the outcome of one determines the outcome of the other. An obvious measure that satisfies these 'boundary conditions' is that of the logarithm of the ratio of the conditional probability $P(x_i | y_j)$ divided by $P(x_i)$, and this defines the mutual information $I(x_i, y_j)$ between x_i and y_j as in Equation 8.18.

$$I(x_i, y_j) = \log \frac{P(x_i | y_j)}{P(x_i)} \tag{8.18}$$

When the random variables are statistically independent, Equation 8.19 and therefore Equation 8.20 are obtained.

$$P(x_i | y_i) = P(x_i) \tag{8.19}$$

$$I(x_i, y_j) = 0 \tag{8.20}$$

However, when the occurrence of event $Y = y_j$ uniquely determines the occurrence of event $X = x_i$, the conditional probability above is unity.

Therefore Equation 8.21 is obtained which is called the self information of the event $X=x_i$.

$$I(x_i, y_j) = \log \frac{1}{P(x_i)} = -\log P(x_i) \tag{8.21}$$

It should be noted that an event that occurs with high probability conveys less information than a low probability event.

176 Mutual and self information

Consider the expression given above for the mutual information, and since we can write Equation 8.22 then the mutual information obeys the symmetry relation given in Equation 2.23 by virtue of Bayes' theorem.

$$\frac{P(x_i|y_j)}{P(x_i)} = \frac{P(x_i|y_j)P(y_j)}{P(x_i)P(y_j)}$$

$$= \frac{P(x_i, y_j)}{P(x_i)P(y_j)} = \frac{P(y_j|x_i)}{P(y_j)} \tag{8.22}$$

$$I(x_i, y_j) = I(y_j, x_i) \tag{8.23}$$

The average value of the mutual information defined above is given by Equation 8.24

$$I(X, Y) = \sum_{i=1}^{n} \sum_{j=1}^{m} P(x_i, y_j) I(x_i, y_j)$$
$$= \sum_{i=1}^{n} \sum_{j=1}^{m} P(x_i, y_j) \log \frac{P(x_i, y_j)}{P(x_i)P(y_j)} \tag{8.24}$$

Likewise, the average value of the self information, $H(X)$ is defined by Equation 8.25.

$$H(X) = \sum_{i=1}^{n} P(x_i) I(x_i)$$
$$= -\sum_{i=1}^{n} P(x_i) \log P(x_i) \tag{8.25}$$

For any given communication source transmitting an 'alphabet' of symbols, the entropy of the source is a maximum when the output symbols are equally probable, as is clear since if the symbols are equally probable, Equations 8.26 and 8.27 and in general Equation 8.28 are obtained.

$$P(x_i) = \frac{1}{n} \quad \text{for all } i \tag{8.26}$$

$$H(X) = -\sum_{i=1}^{n} \frac{1}{n} \log \frac{1}{n} = \log n \tag{8.27}$$

$$H(X) \leq \log n \tag{8.28}$$

It is very straightforward to show that Equation 8.29 follows with equality if and only if X and Y are independent, and that Equation 8.30 is true where $H(X|Y)$ is a conditional uncertainty, etc.

$$H(X, Y) \leq H(X) + H(Y) \tag{8.29}$$

$$H(X, Y) = H(X) + H(Y|X) = H(Y) + H(X|Y) \tag{8.30}$$

The above definitions of mutual information, self information and entropy for discrete random variables may be easily extended to the case of continuous random variables, and if the joint probability density functions and the marginal probability density function are $p(x,y)$, $p(x)$ and $p(y)$ respectively, then the average mutual information between X and Y (or the cross entropy) is given by Equation 8.31 and the entropy of X in the continuous case by Equation 8.32 where $m(x)$ is an appropriate measure function (Jaynes, 1989).

$$I(X, Y) = \int_{-\infty}^{\infty} \int_{-\infty}^{\infty} p(x) p(y|x) \log \frac{p(y|x) P(x)}{p(x) p(y)} dx\, dy \tag{8.31}$$

$$H(X) = -\int_{-\infty}^{\infty} p(x) \log \frac{p(x)}{m(x)} dx \tag{8.32}$$

8.5 Discrete memoryless channels

The discrete memoryless channel serves as a statistical model with an input X and an output Y which is a noisy version of X, and both X and Y are random variables.

The channel is said to be 'discrete' when both the alphabets from which X and Y are selected have finite size, not necessarily the same size, and the channel is said to be 'memoryless' when the current output symbol depends only on the current input symbol and not on any previous ones (c.f. Markov models and ARMA processes etc.).

In terms of the elements of the input and output alphabets, one can define a set of transition probabilities for all j and k, as in Equation 8.33 and this is just the conditional probability that the channel output is $Y = y_k$ given that the channel input is $X = x_j$.

$$p(y_k | x_j) = P(Y = y_k | X = x_j) \tag{8.33}$$

Due to transmission errors, when $k = j$, the transition probability represents the conditional probability of correct reception, and when $k \neq j$ the conditional probability of error. Just as for Markov processes, one can define a transition probability matrix as follows.

$$P = \begin{bmatrix} p(y_0|x_0) & p(y_1|x_0) & \dots & p(y_{K-1}|x_0) \\ p(y_0|x_1) & p(y_1|x_1) & \dots & p(y_{K-1}|x_1) \\ \cdot & \cdot & \dots & \cdot \\ \cdot & \cdot & \dots & \cdot \\ \cdot & \cdot & \dots & \cdot \\ p(y_0|x_{J-1}) & p(y_1|x_{J-1}) & \dots & p(y_{K-1}|x_{J-1}) \end{bmatrix}$$

This is a J by K matrix called the channel matrix. The sum of elements along any row is given by Equation 8.34 for all j.

$$\sum_{k=0}^{K-1} p(y_k | x_j) = 1 \tag{8.34}$$

If the inputs to a discrete memoryless channel are selected according to the probability distribution $p(x_j), j = 0,1,...,J-1$, then the event $X = x_j$ occurs with probability given by Equation 8.35.

$$p(x_j) = P(X = x_j) \tag{8.35}$$

The joint probability distribution of the random variables X and Y is given by Equation 8.36 and the marginal distribution of the output variable Y is obtained by integrating out, or averaging out, the dependence on x_j for $(j = 0,1,...,J-1)$, as in Equation 8.37 for $k = 0,1,...,K-1$.

$$\begin{aligned} p(x_j, y_k) &= P(X = x_j, Y = y_k) \\ &= P(Y = y_k \mid X + x_j) P(X = x_j) \\ &= p(y_k \mid x_j) p(x_j) \end{aligned} \tag{8.36}$$

$$p(y_k) = \sum_{j=0}^{J-1} p(y_k \mid x_j) p(x_j) \tag{8.37}$$

This is a very important equation, since if we are given the input a priori probabilities $p(x_j)$ and the channel matrix, then we may obtain the probabilities of the various output symbols from the above equation.

8.5.1 Channel capacity

Regardless of whatever means, or channel, is used for transmission, there is a maximum rate of transmission, called the 'capacity' of the channel, which is determined by the intrinsic properties of the channel and is independent of the content of the transmitted information and the way it is encoded. This is measured in bits per second. As an example, in order to transmit a colour television picture, a channel with a capacity of about 200 million bits per second is required.

180 Discrete memoryless channels

For a discrete memoryless channel with transition probabilities $p(y_k | x_j)$ as before, the average mutual information between the output and the input is given by Equation 8.38.

$$I(X,Y) = \sum_{j=0}^{J-1} \sum_{k=0}^{K-1} p(x_j, y_k) \log \frac{p(y_k | x_j)}{p(y_k)} \tag{8.38}$$

Also as before, Equations 8.39 and 8.40 may be obtained.

$$p(x_j, y_k) = p(y_k | x_j) p(x_j) \tag{8.39}$$

$$p(y_k) = \sum_{j=0}^{J-1} p(y_k | x_j) p(x_j) \tag{8.40}$$

From the expression for the average mutual information, it is seen therefore that the average mutual information depends both on the channel characteristics expressed in terms of the elements of the channel matrix, and also on the input probability distribution, which is clearly independent of the channel. Therefore, by changing the input probability distribution, the average mutual information will change, and we can define the 'channel capacity' in terms of the maximum average mutual information with respect to the input probability distribution.

We can therefore write the channel capacity as in Equation 8.41.

$$C = \max \ I(X, Y) \tag{8.41}$$

The calculation of C is a constrained optimisation, since the constraints of Equations 8.42 and 8.43 have to apply to $p(x_j)$.

$$p(x_j) = 0 \tag{8.42}$$

$$\sum_{j=1}^{J-1} p(x_j) = 1 \tag{8.43}$$

This may be compared to the method employed for the maximum entropy analysis of images and inverse problems (Jaynes, 1989; Skilling, 1988).

The channel capacity is an extremely important quantity, since it is possible to transmit information through a channel at any rate less than the channel capacity with an arbitrary small probability of error; completely reliable transmission is not possible if the information processed is greater than the channel capacity. However, in general, the calculation of the channel capacity is a difficult problem.

Before calculating the channel capacity for some simple channel models, it is useful to introduce certain classes of channel, as follows:

1. A channel is lossless if $H(X|Y) = 0$ for all input distributions, which means that the input is determined by the output, and hence no transmission errors can occur.
2. A channel is deterministic if $p(y_j|x_i) = 1$ or 0 for all i, j, which means that Y is determined by X, and hence $H(Y|X) = 0$ for all input distributions.
3. A channel is noiseless if it is lossless and deterministic.
4. A channel is useless or zero capacity if $I(X|Y) = 0$ for all input distributions.
5. A channel is symmetric if each row of the transition matrix contains the same set of numbers, and if each column contains the same set of numbers (different in general for the row set).

It is a consequence of the definition of the symmetric channel that $H(Y|X)$ is independent of the input distribution $p(x)$, and depends only on the channel probabilities $p(y_j|x_i)$.

The best known model for a symmetric channel is the binary symmetric channel shown in Figure 8.2. As an example of the calculation of the channel capacity, a closed form expression for a symmetric channel will be given.

182 Discrete memoryless channels

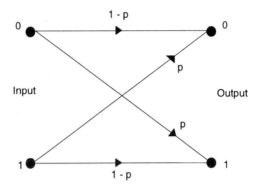

Figure 8.2 Binary symmetric channel

Let us consider a symmetric channel with input alphabet $x_1,...,x_M$, and output alphabet $y_1,...,y_L$, and a channel matrix with row probabilities $p_1',...,p_L'$ and column probabilities $q_1',...,q_M'$. Since $H(Y|X)$ does not depend on the input distribution, the problem of maximising the information reduces to the problem of maximising the output uncertainty $H(Y)$ as in Equation 8.44.

$$I(X|Y) = H(Y) - H(Y|X) \qquad (8.44)$$

It is known that Equation 8.45 holds with equality if and only if all values of Y are equally likely.

$$H(Y) \leq \log L \qquad (8.45)$$

Therefore, if an input distribution can be found for which all values of Y have the same probability, then that input distribution would maximise $I(X|Y)$. The uniform distribution does this, since Equation 8.46 holds and for a uniform distribution Equation 8.47 holds.

$$p(y_j) = \sum_{i=1}^{M} p(x_i) p(y_j|x_i) = \frac{1}{M} \sum_{i=1}^{M} p(y_j|x_i) \qquad (8.46)$$

$$p(x_i) = \frac{1}{M} \text{ for all } i \tag{8.47}$$

However, the term $\sum_{i=1}^{M} p(y_j | x_i)$ is the sum of the entries in the jth column of the channel matrix, and since the channel is symmetric, Equation 8.48 is obtained, which is independent of j, and hence $p(y_j)$ does not depend on j, or equivalently all values of Y have the same probability.

$$\sum_{i=1}^{M} p(y_j | x_i) = \sum_{k=1}^{M} q_k' \tag{8.48}$$

The maximum, given by Equation 8.49, is therefore attainable and the channel capacity can be written as in Equation 8.50 since Equation 8.51 is also true.

$$H(Y) = \log L \tag{8.49}$$

$$C_{sym} = \log L + \sum_{j=1}^{L} p_j' \log p_j' \tag{8.50}$$

$$H(Y|X) = -\sum_{j=1}^{L} p_j' \log p_j' \tag{8.51}$$

For the binary symmetric channel shown in Figure 8.2, the channel capacity can be written as in Equation 8.52 and the variation with β is shown in Figure 8.3.

$$\begin{aligned} C_{BSC} &= \log 2 + \beta \log \beta + (1 - \beta) \log (1 - \beta) \\ &= 1 - H(\beta, 1 - \beta) \end{aligned} \tag{8.52}$$

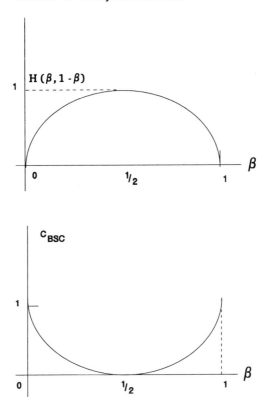

Figure 8.3 Capacity of a binary symmetric channel

8.5.2 Source encoding

An important problem in communications is how to efficiently represent data generated by a discrete source. This is called source encoding. For a source encoder to be efficient, one needs knowledge of the statistics of the source. For example, if some source symbols are known to be more probable than others, then this feature may be exploited by assigning short code words to frequent source symbols and vice versa, e.g. the Morse code.

The average length of a code word, L can be written as in Equation 8.53.

$$L = \sum_{k=1}^{K} p_k l_k \tag{8.53}$$

This represents the average number of bits per source symbol used in the source encoding process. If L_{min} represents the minimum possible value of L, the coding efficiency of the source encoder is defined as in Equation 8.54 and the source encoder is said to be efficient when η approaches unity.

$$\eta = \frac{L_{min}}{L} \tag{8.54}$$

One question of importance is how to determine the minimum length code. The answer to this is given in Shannon's first theorem, also called the source coding theorem. This states that given a discrete memoryless source of entropy $H(X)$, then the average code word length L for any source encoding is bounded as in Equation 8.55.

$$L \geq H(X) \tag{8.55}$$

Therefore, the entropy $H(X)$ represents a fundamental limit on the average number of bits per source symbol necessary to represent a discrete memoryless source. Thus Equation 8.56 can be obtained and the efficiency of the source encoder may be written in terms of the entropy as in Equation 8.57.

$$L_{min} = H(X) \tag{8.56}$$

$$\eta = \frac{H(X)}{L} \tag{8.57}$$

A theorem which is complementary to Shannon's first theorem and applies to a channel in which the noise is Gaussian is known as the

186 Discrete memoryless channels

Shannon-Hartley theorem, which states that the channel capacity of a white, bandlimited Gaussian channel is given by Equation 8.58, where B is the channel bandwidth, S the signal power, and N is the total noise within the channel bandwidth.

$$C = B \log_2\left(1 + \frac{S}{N}\right) \qquad \text{bits/s} \tag{8.58}$$

This theorem, although restricted to Gaussian channels, is of fundamental importance, since many channels are approximately Gaussian, and the results obtained by assuming a Gaussian channel often provide a lower bound on the performance of a system operating over a non-Gaussian channel. The Shannon-Hartley theorem indicates that a noiseless Gaussian channel $\left(\frac{S}{N} = \infty\right)$ has an infinite capacity. However, when noise is present in the channel, the capacity does not approach infinity as the bandwidth is increased, since the noise power increases as the bandwidth increases, and the channel capacity reaches a finite upper limit with increasing bandwidth if the signal power remains constant. This limit is easily calculated using the fact that the total noise within the channel bandwidth is given by Equation 8.59 where $\frac{\eta}{2}$ is the two-sided power spectral density. The Shannon-Hartley theorem may now be written as in Equations 8.60 and 8.61.

$$N = \eta B \tag{8.59}$$

$$\begin{aligned} C &= B \log_2\left(1 + \frac{S}{\eta B}\right) \\ &= \left(\frac{S}{\eta}\right)\left(\frac{\eta B}{S}\right) \log_2\left(1 + \frac{S}{\eta B}\right) \end{aligned} \tag{8.60}$$

$$C = \frac{S}{\eta} \log_2\left(1 + \frac{S}{\eta B}\right)^{\eta B/S} \tag{8.61}$$

Using Equation 8.62 gives Equation 8.63.

$$x = \frac{S}{\eta B} \tag{8.62}$$

$$\lim_{B \to \infty} C = \frac{S}{\eta} \log_2 e = 1.44 \frac{S}{\eta} \tag{8.63}$$

The Shannon-Hartley theorem allows for a trade off between bandwidth and signal to noise ratio, in the following sense. If, for example, the signal to noise ratio $S/N = 7$, and $B = 4\text{kHz}$, we obtain $C = 12 \times 10^3$ bits/s. If the signal to noise ratio is increased to 15, and the bandwidth increased to 3kHz, the channel capacity remains the same.

We have seen that in the case of a noiseless channel, the capacity is infinite, and no matter how restricted the bandwidth, it is always possible to receive a signal without error, as predicted by the Shannon limit. The question of how in practice one achieves this performance is complex, one method being the use of orthogonal signals.

Let us now consider the problem of the reliable transmission of messages through a noisy communications channel. We therefore need the best decoding scheme to enable us to obtain the input sequence after seeing the received symbols.

Assuming the input alphabet $x_1,...,x_M$, output alphabet $y_1,...,y_L$, and channel matrix $p(y_j | x_i)$ as before, and consider the special case where a sequence of symbols, chosen at random according to a known distribution, $p(x)$, is transmitted through the channel. For this given distribution $p(x)$, we wish to construct a decision scheme that minimises the overall probability of error. Such a scheme is called an ideal observer.

Consider a sequence $x = (\alpha_1, ..., \alpha_n)$ chosen in accordance with the distribution $p(x)$. The probability that the output sequence $\beta_1, ..., \beta_n$ is produced is given by Equation 8.64, which may also be written as Equation 8.65.

$$p(\beta_1, ..., \beta_n = \sum_{\alpha_1,...,\alpha_n} p(\alpha_1,...,\alpha_n) \\ \times p(\beta_1,...,\beta_n | \alpha_1,...,\alpha_n) \tag{8.64}$$

$$p(\beta_1,...,\beta_n) = \sum_{\alpha_1,...,\beta_n} p(\alpha_1,...,\alpha_n)$$
$$\times p(\beta_1 | \alpha_1 (p(\beta_2 | \alpha_2) ... p(\beta_n | \alpha_n))) \quad (8.65)$$

A decoder scheme may be defined as a function that assigns to each output sequence ($\beta_1,...,\beta_n$) an input sequence ($\alpha_1,...,\alpha_n$) and the ideal observer is the scheme which minimises the overall probability of error for the given input distribution, and this is found by maximising the conditional probability as given by Equation 8.66.

$$p(\alpha_1,...,\alpha_n | \beta_1,...,\beta_n)$$
$$= \frac{p(\alpha_1,...,\alpha_n) \prod_{k=1}^{n} p(\beta_k | \alpha_k)}{p(\beta_1,...,\beta_n)} \quad (8.66)$$

When all the inputs are equally likely, then Equation 8.67 may be obtained and for a fixed y, maximising $p(x_i | y)$ is equivalent to maximising the inverse probability $p(y | x_i)$. This is referred to as the maximum likelihood decision scheme.

$$p(x_i | y) = \frac{p(x_i) p(y | x_i)}{p(y)} = \frac{1}{M p(y)} p(y | x_i) \quad (8.67)$$

8.6 Discrete channels with memory

Previously we have described channels that have no memory, that is channels in which the occurrence of errors during a particular symbol interval does not influence the occurrence of errors during succeeding symbol intervals. However in many realistic channels, errors do not occur as independent random events, but tend to occur in bursts. These channels are said to have memory. Examples of channels with memory would be telephone channels that suffer from switching transients, and microwave radio links that suffer from fading, etc.

It is useful to give two simple models for discrete channels with memory, before going onto ideas associated with capacity and coding for these channels. The first model is due to Blackwell (1961), and

Information theory 189

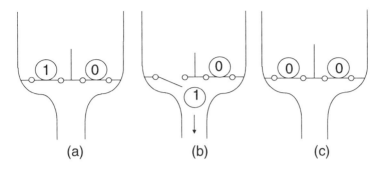

Figure 8.4 Trapdoor channel

considers 'trapdoors' as shown in Figure 8.4. Initially a ball labelled with 0 or 1 is placed in each of the two slots. One of the trapdoors is then opened, with each door having the same probability of being opened. The ball then falls through the open door, and the door then closes behind it. The empty compartment then has another ball placed in it, and the whole process starts all over again. This model defines a channel whose inputs correspond to the balls placed in the empty compartment and whose outputs correspond to the balls which fall through the trapdoors. If the symbol b_i corresponds to the condition in which a ball labelled i remains in an occupied slot, and time is started after one of the doors is opened, four states s_{ij}, for $i,j = 0$ or 1 may be defined as follows. The channel is in state s_{ij} at time $t = n$, if the condition b_j holds at time $t = n$ and condition b_i hold at time $t = n-1$. An input k therefore corresponds to the placing of a ball labelled k in the unoccupied slot, and the opening of the trapdoor then determines the corresponding output and the next state.

If at time $t = n$ the channel is in state s_{10}, and an input 1 is applied, then one ball labelled 0 and one ball labelled 1 rest over the trapdoors. With a probability of 1/2 the ball 1 falls through, leaving the ball 0 in the occupied slot. The channel then moves to state s_{00} and emits an output of 1. With probability 1/2 the 0 ball falls through, sending the channel into the state s_{01}, and an output 0 is emitted. The behaviour of this channel may thus be described by two matrices, M_0 and M_1, input = 0 $\quad M_0 = \begin{array}{c} \\ S_{00} \\ S_{10} \\ S_{01} \\ S_{11} \end{array} \begin{array}{cccc} S_{00} & S_{10} & S_{01} & S_{11} \\ \left[\begin{array}{cccc} 1 & 0 & 0 & 0 \\ 1 & 0 & 0 & 0 \\ 0 & \frac{1}{2} & 0 & \frac{1}{2} \\ 0 & \frac{1}{2} & 0 & \frac{1}{2} \end{array}\right] \end{array}$

input = 1 $\quad M_1 = \begin{array}{c} \\ S_{00} \\ S_{10} \\ S_{01} \\ S_{11} \end{array} \begin{array}{cccc} S_{00} & S_{10} & S_{01} & S_{11} \\ \left[\begin{array}{cccc} \frac{1}{2} & 0 & \frac{1}{2} & 0 \\ \frac{1}{2} & 0 & \frac{1}{2} & 0 \\ 0 & 0 & 0 & 1 \\ 0 & 0 & 0 & 1 \end{array}\right] \end{array}$

Figure 8.5 Channel martices for the trapdoor channel

called the channel matrices, whose components are the state transition probabilities under the input 0 and 1 respectively, and a function that associates an output with each input-state pair, Figure 8.5.

Another model that is successful in characterising error bursts and impulsive noise in channels is the so called Gilbert model, where the channel is modelled as a discrete memoryless binary symmetric channel for which the probability of error is a time varying parameter. The changes in probability of error are modelled as a Markov process as shown in Figure 8.6. The error generating mechanism in the channel occupies one of the three states. When the channel is in state 2, the probability of error during a bit interval is 10^{-2} and the channel stays in this state during the succeeding bit interval with a probability of 0.998. The channel may make a transition to state 1, which has a bit error probability of 0.5, and since the system stays in this state with probability 0.99, errors tend to occur in groups.

The models developed for the case of a discrete channel with memory are those for which the channel has a finite number of internal states, for which the present state of the channel represents a summary of its past history. The application of an input will result in a transition to another state, with the production of an output. The

Discrete channels with memory 191

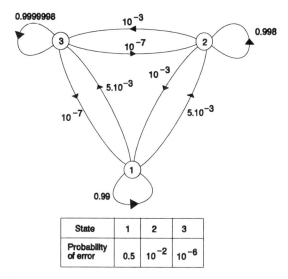

Figure 8.6 A three state Gilbert model

resulting source channel matrix then determines a finite Markov chain with states (a_i, s_k) where a_i are the states of the source and s_k are the states of the channel. If the chain associated with the source channel matrix has steady state probabilities, then given the pair (a_i, s_k) one can determine the corresponding input $f_1(a_i, s_k) = f(a_i)$ and output $g_1(a_i, s_k) = g(f(a_i), s_k)$. Thus the chain associated with the source channel matrix determines a stationary sequence of input-output pairs (X_n, Y_n), for $n = 1, 2, \ldots$. An input, output and joint uncertainty may therefore be defined as in Equations 8.68 to 8.70.

$$H(X) = \lim_{n \to \infty} H(X_n \mid X_1, \ldots X_{n-1}) \qquad (8.68)$$

$$H(Y) = \lim_{n \to \infty} H(Y_n \mid Y_1, \ldots Y_{n-1}) \qquad (8.69)$$

$$H(X, Y) = \lim_{n \to \infty} H[(X_n, Y_n) \mid (X_1, Y_1), \ldots (X_{n-1}, Y_{n-1})] \qquad (8.70)$$

The information conveyed about the process X by the process Y is given by Equation 8.71.

$$I[X|Y] = H(X) + H(Y) - H(X,Y) = I[Y|X] \qquad (8.71)$$

The capacity of the given channel is defined as the least upper bound of $I[X|Y]$, taken over all regular Markov sources, and it can be proved that it is possible to transmit information at any rate less than the capacity with an arbitrarily small probability of error.

The above discussion says nothing about how one calculates the channel capacity of a given finite-state channel. This is an extremely complex task, and details may be found in Proakis (1987).

It is possible to model a channel such that the model combines features of the discrete memoryless channel with those of the discrete channel with memory, and this defines a compound channel.

8.7 Continuous channels

Communications channels may be continuous in two senses. In the first sense, we may allow the input and output alphabets to contain an infinite number of elements, but we require that the transmitted material be in the form of a discrete sequence of symbols. This type of channel is called a time-discrete, amplitude continuous channel.

For the second type of continuous channel, we allow the transmission of information to be continuous in time.

Consider an analog source that emits a message waveform $x(t)$ which is a sample function of a stochastic process $X(t)$. If $X(t)$ is a stationary stochastic process with autocorrelation function $\Phi_{xx}(\tau)$ and power spectral density function $\Phi_{xx}(f)$, and if $X(t)$ is a band-limited process such that Equation 8.72 is satisfied, then $X(t)$ has a representation given by the sampling theorem as in Equation 8.73.

$$\Phi_{xx}(f) = 0 \quad \text{for } |f| > W \qquad (8.72)$$

$$X(t) = \sum_{n=-\infty}^{\infty} X\left(\frac{n}{2W}\right) \frac{\sin 2\pi W\left(t - \frac{n}{2W}\right)}{2\pi W\left(t - \frac{n}{2W}\right)} \qquad (8.73)$$

This means that the bandlimited signal can be represented by a sequence of samples as in Equation 8.74 sampled at the rate given by Equation 8.75 the so called Nyquist rate. These samples may then be encoded using various techniques giving a digital representation of the analog signal.

$$X_n = X\left(\frac{n}{2W}\right) \qquad (8.74)$$

$$f_s = 2W \quad \text{samples per second} \qquad (8.75)$$

One such technique is Pulse Code Modulation (PCM), the essential operations being sampling, quantising and encoding, usually performed in the same circuit, called an analog-to-digital converter (ADC). It should be noted that PCM is not modulation in the conventional sense of the word, since modulation usually refers to the variation of some characteristic of the carrier wave in accordance with an information bearing signal. The only part of PCM that is similar to this definition, is sampling. The subsequent use of quantisation, which is basic to PCM, introduces a signal distortion that has no counterpart in conventional modulation.

Let $x(t)$ denote a sample function emitted by a source, and let x_n represent the samples taken at a sampling rate greater than the Nyquist rate, $2W$, where W is the highest frequency present in the signal. In PCM, each sample is quantised to one of 2^b amplitude levels, where b is the number of binary digits used to represent each sample. The rate from the source is therefore bf_s bits per second, where f_s is the sampling frequency.

The quantisation may be represented as in Equation 8.76, where $\tilde{x}_n$ represents the quantised value of x_n and q_n represents the quantisation error which is treated as additive noise.

194 Information theory

$$\tilde{x}_n = x_n + q_n \tag{8.76}$$

If an uniform quantiser is used, the quantisation noise is well represented by a uniform probability density function, and the mean square value of the quantisation error can be shown to be as in Equation 8.77, where $\Delta = 2^{-b}$ is the step size of the quantiser.

$$E(q^2) = \frac{\Delta^2}{12} = \frac{2^{-2b}}{12} \tag{8.77}$$

Measured in decibels, the mean square value of the noise is as in Equation 8.78.

$$10 \log \frac{\Delta^2}{12} = -6b - 10.8 \; dB \tag{8.78}$$

For example, a 7-bit quantiser gives a quantisation noise power of -52.8 dB.

A uniform quantiser provides the same spacing between successive levels throughout the entire dynamic range of the signal. In practice this can cause problems since for many signals, speech for example, small signal amplitudes occur more frequently than large signal amplitudes, and a better approach would have more closely spaced levels at low signal amplitudes, and vice versa. The desired form of nonuniform quantisation can be achieved by using a compressor followed by a uniform quantiser. By cascading this combination with an expander complementary to the compressor, the original signal samples are restored to their correct values apart for quantisation errors. This combination of compressor and expander is called a compander.

There are basically two types of compander in current use, μ-law and A-law.

The transfer characteristic of the compressor is represented by a memoryless nonlinearity $c(x)$, where x is the sample value of a random variable X denoting the compressor input. In the μ-law compander, $c(x)$ is continuous, approximating a linear dependence on

x for low input levels and a logarithmic one for high input levels, as in Equation 8.79 for $0 \leq \frac{|x|}{x_{max}} \leq 1$.

$$\frac{c(|x|)}{x_{max}} = \frac{\ln\left(1 + \frac{\mu|x|}{x_{max}}\right)}{\ln(1+\mu)} \qquad (8.79)$$

The special case of uniform quantisation corresponds to the case where $\mu = 0$. The μ-law is used for PCM telephone systems in the USA, Canada and Japan.

In A-law companding, the compressor characteristic is piecewise, made up of a linear segment for low-level inputs and a logarithmic segment for high-level inputs. This type of compander is used for the PCM telephone systems in Europe.

In PCM each sample of the waveform is encoded independently of all the other samples. However, most source signals sampled at or above the Nyquist rate exhibit significant correlation between successive samples, i.e. the average change in amplitude between successive samples is relatively small. An encoding scheme that exploits the redundancy in the samples will result in a lower bit rate for the source output, and one such system called Differential pulse code modulation (DPCM) does this by encoding the differences between successive samples rather than the samples themselves.

A sub-class of DPCM is delta modulation in which the code word is only one binary digit in length. Delta modulators have the advantage over more conventional PCM systems of having a particularly simple architecture, which forms the basis of 'single-chip' coder-decoder (CODEC) integrated circuits. Because delta modulation is a 1-digit code, sampling rate and output digit rate are the same. However, in order for delta modulation to achieve performance comparable to that of conventional n-digit PCM systems, the sampling rate must be increased substantially over that required for a PCM system.

Most real signal sources are nonstationary in nature, which means that the variance and the autocorrelation function of the source output vary with time. PCM and DPCM encoders are designed on the basis that the source output is stationary, and the performance of these encoders can be improved by using adaptive methods. One such method being the use of an adaptive quantiser.

PCM, DPCM and the adaptive counterparts are source encoding techniques that attempt to represent the output waveforms. Consequently these methods are known as waveform encoding techniques. In contrast to these methods, Linear Predictive Coding models the source as a linear system, which when excited by an appropriate input signal gives rise to an observed source output. Instead of transmitting the samples of the source waveform to the receiver, the model parameters of the linear system are transmitted together with the appropriate excitation signal.

8.8 References

Cox, R.T. (1946) *Am.J.Phys.* 17.1.1946.

Gull, S. (1988) Developments in Maximum Entropy data analysis. *In Maximum Entropy and Bayesian Methods* (ed. J. Skilling). Kluwer Academic Press.

Hartley, R.V.L. (1928) Transmission of Information. *Bell Systems Technical Journal*, July.

Jaynes, E.T. (1978) Where do we stand on Maximum Entropy. *In The Maximum Entropy Formulism* (ed. R.D. Levine and M. Tribus) M.I.T. Press.

Jaynes, E.T. (1989) *Papers on probability, statistics and statistical physics* (ed. R.D. Rosenkranz) Kluwer Academic Press, 1989.

Leff, H.S. and Rex, A.F. (1990) *Maxwell's Demon, Entropy, Information, Computing*. Adam Hilger Press.

Proakis, J. (1987) *Digital Communications*. McGraw-Hill.

Shannon, C. (1949) *The Mathematical Theory of Communication*, University of Illinois Press.

Skilling, J. (ed.) (1988) *Maximum Entropy and Bayesian Methods*. Kluwer Academic Press.

9. Teletraffic theory

9.1 Introduction

The subject of Teletraffic Theory has a long history dating back to the work of the Danish mathematician, A.K. Erlang (Brockmeyer et al., 1948), first published in 1917.

Today the subject has grown enormously and is still growing at a rapid pace with research into the performance modelling of areas as diverse as, for example, dynamic routeing strategies, cellular mobile radio systems and the application of fast packet switching techniques proposed for use in a broadband integrated services digital network. Hui (1990) is a good text for some of these more recent developments.

In Section 9.2 the basic model of a single link in a circuit switched telecommunication network is studied. This section also introduces the important control mechanism of trunk reservation.

Section 9.3 shows how the single link models studied in Section 9.2 can be incorporated into models of networks of links using the effective link independence approximation to construct the Erlang fixed point systems of equations.

These equations allow the investigation of the rich behaviour of dynamic routeing strategies in the setting of symmetric fully connected networks. This section on network based results ends with two bounds on the overall best possible behaviour of dynamic routeing strategies.

Section 9.4 presents some of the dynamic routeing strategies that have been either implemented or proposed for real networks.

This is an area which has received much attention over the last decade and is one in which teletraffic theory has provided the models and understanding leading to more efficient, flexible and reliable networks.

9.2 Single link models

9.2.1 Erlang's loss formula

The first model to be considered is the classical model for a single link in a circuit switched network. Suppose that the arrival process of calls offered to a link is a Poisson process of rate λ. Let the holding times of calls be exponentially distributed with a mean of 1 independently of the arrival process. Suppose that the link has a capacity of C circuits. See Figure 9.1 and Chapter 7 for an introduction to the theory of stochastic processes applied to queuing networks.

This system may be modelled by the stochastic process, n, the number of circuits occupied by the link with n taking values in the range 0 to C inclusive. It can be seen that n is a birth and death process since n either increases by one when a new call is accepted on the link or decreases by one when an existing call clears down. The transition rates for the birth and death process are therefore given by Equations 9.1 and 9.2.

$$q(n, n+1) = \lambda \quad n = 0, 1, ..., C-1 \tag{9.1}$$

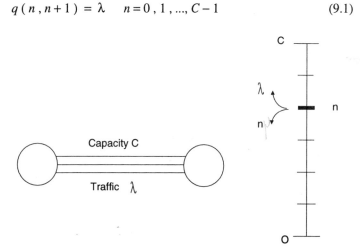

Figure 9.1 Single link model

$$q(n, n-1) = n \quad n = 1, ..., C \tag{9.2}$$

A general result, (Kelly, 1979) on birth and death processes gives the equilibrium distribution in terms of the birth and death rates, as in Equation 9.3 where π_0 is determined from the normalisation condition of Equation 9.4.

$$\pi_n = \pi_0 \prod_{i=1}^{n} \frac{q(i-1, i)}{q(i, i-1)} \tag{9.3}$$

$$\sum_{n=0}^{C} \pi_n = 1 \tag{9.4}$$

Hence, from Equations 9.1 and 9.2 one can obtain Equations 9.5 and 9.6.

$$\pi_n = \pi_0 \prod_{i=1}^{n} \frac{\lambda}{i} = \pi_0 \frac{\lambda^n}{n!} \tag{9.5}$$

$$\pi_0 = \left[\sum_{n=0}^{C} \frac{\lambda^n}{n!} \right]^{-1} \tag{9.6}$$

In summary, the equilibrium distribution for the utilisation, n, of a single link is given by Equation 9.7.

$$\pi_n = \frac{\dfrac{\lambda^n}{n!}}{\displaystyle\sum_{i=0}^{C} \frac{\lambda^i}{i!}} \quad (n = 0, 1, ..., C) \tag{9.7}$$

200 Single link models

In particular, the probability, L, that an arriving call finds the link full to capacity is π_C given by Equation 9.8.

$$L = \pi_C = \frac{\lambda^C}{C!} \left[\sum_{i=0}^{C} \frac{\lambda^i}{i!} \right]^{-1} \tag{9.8}$$

Equation 9.8 is known as Erlang's loss formula and is undoubtedly the most widely used result in teletraffic theory. The standard notation $E(\lambda,C)$ for the quantity given in Equation 9.9 will be used.

$$E(\lambda, C) = \frac{\lambda^C}{C!} \left[\sum_{i=0}^{C} \frac{\lambda^i}{i!} \right]^{-1} \tag{9.9}$$

Later the question of how best to compute $E(lambda,C)$ in such a way as to reduce numerical inaccuracies caused by rounding errors will be considered. First, let us look at some of its properties.

In Figure 9.2 the loss probability $L = E(\lambda,C)$ is shown as a function of the traffic intensity $\rho = \lambda/C$ for C held fixed at the values $C = 100, 250, 1000$ and 10000B. This figure demonstrates the important trunking efficiency effect where for a fixed level of the traffic intensity, ρ, the loss probability, L, decreases with increasing link capacity, C: larger links are more efficient at carrying calls. Also shown in the figure is the limiting case as C increases to infinity which is given by Equation 9.10.

$$L = \max\left(1 - \frac{1}{\rho}, 0\right) \tag{9.10}$$

9.2.1.1 *Numerical considerations*

Several methods are available for the efficient and accurate numerical evaluation of Erlang's loss formula. For small values of C, perhaps up to several hundreds of circuits, $E(\lambda,C)$ can be computed using the recursive formula of Equation 9.11 and $(\lambda,0) = 1$.

Teletraffic theory 201

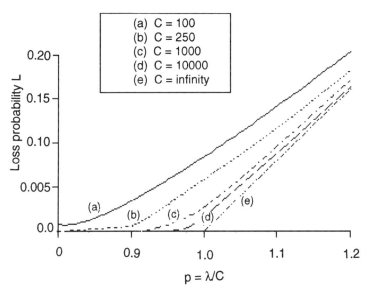

Figure 9.2 Erlang's blocking formula

$$E(\lambda, C) = \frac{\lambda E(\lambda, C-1)}{C + \lambda E(\lambda, C-1)} \tag{9.11}$$

This result can be derived by substituting the expression for $E(\lambda,C)$ given in Equation 9.9 and simplifying the resulting expression.

For larger values of C, which make a recursive computation impractical, a second method is available which avoids the direct calculation of large factorials. After some re-arrangement Equation 9.12 may be obtained.

$$E(\lambda, C) = \left[1 + \frac{C}{\lambda} + \frac{C(C-1)}{\lambda^2} + \ldots + \frac{C!}{\lambda^C} \right]^{-1} \tag{9.12}$$

Written in this form each successive term in the summation is a simple multiple of the previous one and so direct calculation of factorials may be avoided. The final sum is then inverted to give the loss probability.

9.2.2 Multiple priorities and trunk reservation

Consider again a single link with C circuits but now suppose that there are two independent Poisson arrival streams offered at rates λ_1 and λ_2. Suppose we wish to give priority to the calls at rate λ_1 over those at rate λ_2. A very simple mechanism has been devised to implement these priorities known as trunk reservation. In this mechanism, calls of the low priority stream are only accepted when there are at least $r+1$ free circuits available. In contrast, the high priority calls are accepted so long as there is at least 1 free circuit available. The integer value, r, is known as the trunk reservation parameter. An equivalent notation which is sometimes more convenient is to let C^i be the capacity available to calls of priority i or lower so that $C^1 = C$ and $C^2 = C - r$.

Again, a birth and death process can be used to model the system. The transition rates are now given by Equations 9.13 and 9.14.

$$q(n, n+1) = \begin{cases} \lambda_1 + \lambda_2 & n = 0, \ldots, C-r-1 \\ \lambda_1 & n = C-r, \ldots, C \end{cases} \quad (9.13)$$

$$q(n, n-1) = n \quad n = 1, \ldots, C \quad (9.14)$$

The equilibrium distribution, using the general result in Equation 9.3 is given, after some re-arrangement, by Equation 9.15, where π_0 is determined by Equation 9.16.

$$\frac{\pi_n}{\pi_0} = \begin{cases} \dfrac{(C-r)!}{n!} (\lambda_1 + \lambda_2)^{n-(C-r)} & n = 0, \ldots, C-r \\ \dfrac{(C-r)!}{n!} \lambda_1^{n-(C-r)} & n = C-r+1, \ldots, C \end{cases} \quad (9.15)$$

$$\sum_{n=0}^{C} \pi_n = 1 \tag{9.16}$$

The above procedure avoiding large factorials in Erlang's loss formula can also be used here to calculate the distribution π_n.

Two criteria for selecting the trunk reservation parameter r will now be considered.

9.2.2.1 *Optimality criterion for r*

Consider the situation given by Equations 9.17 and 9.18.

$$B_1(r) = \pi_C \tag{9.17}$$

$$B_2(r) = \sum_{k=C-r}^{C} \pi_k \tag{9.18}$$

Then $B_1(r)$ and $B_2(r)$ are the blocking probabilities to the high and low priority streams respectively when the trunk reservation parameter level is set at the value r.

A criterion for choosing r, which is applicable in much of the work to follow on control in routeing strategies in fully connected networks, is to minimise with respect to r the expression of 9.19.

$$\lambda_1 B_1(r) + \frac{1}{2} \lambda_2 B_2(r) \tag{9.19}$$

This would correspond to a valuation by the link that the carrying of a low priority call is worth just half the carrying of a high priority call. Here we think of high priority calls as the directly routed calls and the low priority calls as the overflow calls using two links.

In Figure 9.3 the value of r minimising expression 9.19 is illustrated with λ_1 held fixed in such a way that $E(\lambda,C) = 1\%$ for two values of C.

204 Single link models

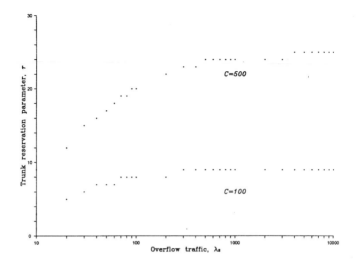

Figure 9.3 Optimality criterion

The value of expression (9.19) is relatively insensitive to r above a low threshold. Thus, while minimising the value of r can be quite sensitive to the precise values of C, λ_1, and λ_2 close to optimal performance over a wide range of values of C, λ_1 and λ_2 can be obtained by quite a crude choice of the parameter r.

9.2.2.2 Secondary criterion for r

A further criterion which may be applied to selecting the trunk reservation parameter r is as follows. The approach begins by first recognising that the offered traffics in a network are in practice very uncertain and large low priority overflow traffics may occur. Trunk reservation may then be used to guarantee performance in the worst case conditions of infinite overflow traffic, where the number of circuits in use on a link is forced to remain in the states $C-r, ..., C$, where $r>0$ is the trunk reservation parameter. Then the transition rates for the birth and death process are simply as in Equations 9.20 and 9.21.

$$q(n, n+1) = \lambda \quad n = C-r, \ldots, C-1 \tag{9.20}$$

$$q(n, n-1) = n \quad n = C-r+1, \ldots, C \tag{9.21}$$

If $B(\lambda, C, r)$ is the blocking probability for fresh traffic under these circumstances then Equation 9.22 may be obtained.

$$B(\lambda, C, r) = \left[\sum_{n=C-r}^{C} \frac{C(C-!)\ldots(n+1)}{\lambda^{C-n}} \right]^{-1} \tag{9.22}$$

A rationale for choosing the trunk reservation parameter is then to relate λ and C by Equation 9.23, where $B > 0$ is a fixed constant, and to take the trunk reservation parameter, $R(C)$, as in Equation 9.24 for some constant K with $1 < K < 1/B$.

$$E(\lambda, C) = B \tag{9.23}$$

$$R(C) = \min\{r : B(\lambda, C, r) \leq KB\} \tag{9.24}$$

Suppose that fresh traffic alone would suffer a blocking probability of B. Then the trunk reservation mechanism with parameter $r = R(C)$ chosen in accordance with this criterion ensures that the blocking probability for fresh traffic under arbitrary conditions of overflow traffic is no worse than K times that without overflow traffic. Figure 9.4 gives some examples of $R(C)$ with several values of the parameter K.

9.3 Network models

9.3.1 Erlang fixed point

In section 9.2 models of a single link were described. In this section the scope will be broadened to investigate models of networks of links. The simplest, and most commonly used, approach to creating a model of a network of links is to use an approximation that blockings

206 Network models

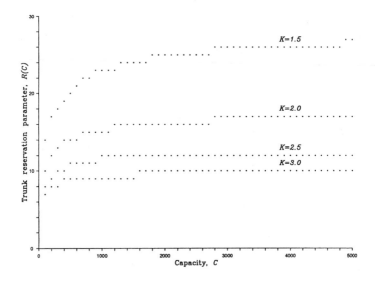

Figure 9.4 Secondary criterion

on links within a route are statistically independent. In order to describe this approximation consider a simple situation of a network of nodes fully connected by links each of capacity C and with independent Poisson arrival streams of traffic offered between the pairs of nodes each of rate λ.

The fully connected network architecture has been the focus of much attention in recent years in relation to studies of dynamic routeing strategies for national trunk networks. Section 9.4 considers in some depth the various strategies that have so far been proposed. In this section a simple strategy for symmetric fully connected networks is considered, that will suffice to illustrate many of the modelling aspects. In practice, networks are not symmetric, and in particular traffics may not be well matched to capacity. Nevertheless, one can learn much about the behaviour of routeing schemes from a consideration of the symmetric case (see Gibbens and Kelly, 1990, for further discussion on dynamic routeing strategies).

Teletraffic theory 207

This simple strategy is called random routeing and operates as follows: a call arriving at the network from the source destination pair (i,j) is routed along link (i,j) if there is at least one free circuit on this link; otherwise an intermediate node $k(k \neq \sim i, j)$ (sometimes called a tandem node) is chosen at random and the call is routed along the two link path between nodes i and j via node k provided that there are at least $r+1$ free circuits on each of the links (i,k) and (k,j); otherwise the call is lost. Notice that each link receives direct and overflow traffic. (See Figure 9.5.) Let B_1 and B_2 be the blocking probabilities of a link for direct and overflow traffic respectively. Observe that B_1 and B_2 will differ as a result of the use of the trunk reservation mechanism with parameter r.

Consider a single link in such a network and let, n, be the number of circuits in use. The process, n, can be modelled by a birth and death process with transition rates as in Equations 9.25 and 9.26, where σ is the rate of overflow traffic.

$$q(n, n+1) = \begin{cases} \lambda + \sigma & n = 0, ..., C - r - 1 \\ \lambda & n = C - r, ..., C - 1 \end{cases} \quad (9.25)$$

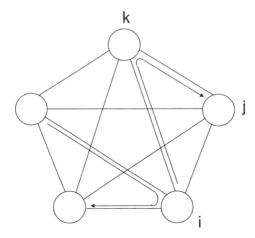

Figure 9.5 A fully connected network

$$q(n, n-1) = n \quad n = 1, ..., C \tag{9.26}$$

The independent blocking approximation allows σ to be expressed in terms of B_1 and B_2 by Equation 9.27.

$$\sigma = 2 \lambda B_1 (1 - B_2) \tag{9.27}$$

The interpretation of Equation 9.27 is that blocking conditions of different links are approximately independent, and that overflow traffic arriving at a link is approximately Poisson. A call blocked with probability B_1 on its direct route will attempt an alternative route through two links, but may be blocked at either of these links. The transition rates correspond to the acceptance by a link of direct traffic and overflow traffic (at rates λ and σ respectively) in states $\{0, ..., C-r-1\}$ and of direct traffic only (at rate λ) in states $\{C-r, ..., C-1\}$.

If L is the probability that a call offered to the network is lost then Equation 9.28 may be obtained.

$$1 - L = (1 - B_1) + B_1 (1 - B_2)^2 \tag{9.28}$$

Since calls are first offered to the direct links and are accepted with probability $(1 - B_1)$ and with probability B_1 they overflow to a randomly chosen two-link alternative path on each link of which they are independently accepted with probability $(1 - B_2)$.

The task is now to solve for B_1 and B_2. Let $\{\pi_n : n = 0, ..., C\}$ be the stationary distribution of the process n then Equations 9.29 and 9.30 are obtained.

$$B_1 = \pi_C \tag{9.29}$$

$$B_2 = \sum_{n=C-r}^{C} \pi_n \tag{9.30}$$

The distribution $\{\pi_n : n = 0, ..., C\}$ is itself a function of B_1 and B_2 and so Equations 9.25 to 9.30 define B_1 and B_2 implicitly as in Equations 9.31 and 9.32 for some functions φ_1 and φ_2.

$$B_1^{I+1} = \varphi_1(B_1^I, B_2^I) \tag{9.31}$$

$$B_2^{I+1} = \varphi_2(B_1^I, B_2^I) \tag{9.32}$$

Choosing initial starting values for B_1^0 and B_2^0 one can use repeated substitution in Equations 9.31 and 9.32 to readily obtain a solution. By the Brouwer fixed point theorem there exists a solution, and indeed as will be seen later there may be more than one solution.

Now, suppose that a call blocked on its direct route is allowed to attempt not just one but up to M two-link alternatives before it is lost. Again, use trunk reservation to insist that at least $r + 1$ circuits be free on a link before it accepts an alternatively routed call. Then the process, n, modelling the utilization of a link is again a birth and death process with rates as given by Equations 9.25 and 9.26 but now with the overflow rate (σ) given by Equation 9.33 and the loss probability (L) given by Equation 9.34.

$$\sigma = 2\lambda B_1 (1 - B_2)^{-1} \left\{ 1 - [1 - (1 - B_2)^2]^M \right\} \tag{9.33}$$

$$L = B_1 [1 - (1 - B_2)^2]^M \tag{9.34}$$

These expressions reduce to those in Equations 9.27 and 9.28 with the choice $M = 1$. The repeated substitution method can be used to determine a solution for B_1 and B_2 given some initial starting values.

Figure 9.6 shows solutions to these fixed point equations when $r = 0$ (i.e. no trunk reservation is applied) with $M = 1, 5$ and a range of values of λ and C. Note that when $r = 0$ we have that $B_1 = B_2 = B$, say. Observe the possibility of multiple solutions for B for C large enough and for a narrow range of the ratio λ/C and that these effects are magnified in the case of larger M. The upper and lower solutions

210 Network models

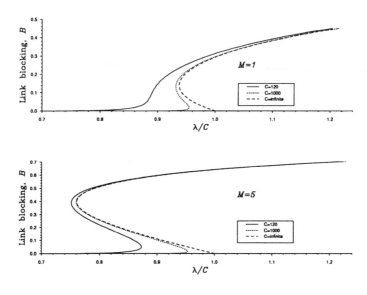

Figure 9.6 Blocking probability B with M = 1.5

correspond to stable fixed point solutions while the middle solution corresponds to an unstable fixed point.

Simulations also exhibit this bistable behaviour where the network has two quasi-stable states. A high blocking state corresponding to many calls being carried on two-link routes and a low blocking state corresponding to only a few calls carried over two links with the majority carried on one-link routes.

Trunk reservation has been found to be an ideal way of preventing this bistable behaviour in the network performance and for providing more efficient use of network resources. In Figure 9.7 the loss probability, L, is shown as a function of r for values $M = 1$ and 5, $\lambda = 110$, and $C = 120$. Observe that a value of r in a broad range will give close to optimal performance.

Now consider the effect of increasing the number of retries M. Figure 9.8 shows the loss probability for varying λ and $C = 120$ and for various values of the retry parameter M. The curves are obtained by choosing the optimal choice of r for each traffic (as in Figure 9.7).

Teletraffic theory 211

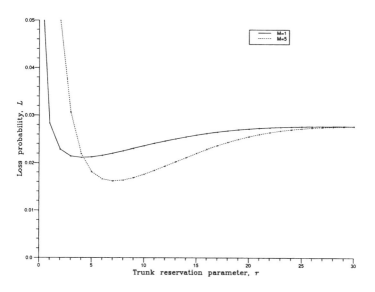

Figure 9.7 Optimal choice of r

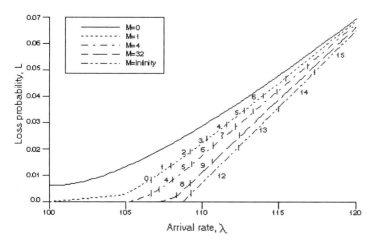

Figure 9.8 Loss probabilities with varying number of retries, M

The optimal choices of r are also shown in Figure 9.8 together with the limiting case where M tends to infinity.

9.3.2 Implied costs

Now examine the issue of implied costs (see Kelly, 1990) in the context of fixed routeing strategies. Suppose that the circuit switched network consists of links labelled $k = 1,2,...,K$ with link k comprising C_k circuits. A subset $r \subset \{1, 2, ..., K\}$ identifies a route. Calls requesting route r arrive as independent Poisson processes of rate v_r. A call requesting route r is blocked and lost if on any link $k \in r$ there are no free circuits. Otherwise, the call is connected and simultaneously holds one circuit on each link $k \in r$ for the holding period of the call. The call holding period is independent of earlier arrival times and holding periods; holding periods of calls on route r are identically distributed with unit mean. Write R for the set of possible routes.

The Erlang fixed point equations for this set up are as in Equation 9.35 and the proportion of calls requesting route r that are lost is given by Equation 9.36.

$$B_k = E\left(\sum_{r:k \in r} v_r \prod_{j \in r-\{k\}} (1-B_j), C_k \right) \tag{9.35}$$

$$L_r = 1 - \prod_{k \in r} (1-B_k) \tag{9.36}$$

Suppose that a call accepted on route r generates an expected revenue ω_r. The rate of return from the network will be as in Equation 9.37 where λ_r is given by Equation 9.38.

$$W(v;C) = \sum_{r \in R} \omega_r \lambda_r \tag{9.37}$$

$$\lambda_r = v_r \prod_{k \in r} (1-B_k) \tag{9.38}$$

Let $c = (c_1, c_2, ..., c_k)$ be the unique solution of the systems of equations given by Equation 9.39.

$$c_k = \eta_k (1 - B_k)^{-1} \sum_{r:k \in r} \lambda_r \left(\omega_r - \sum_{j \in r - \{k\}} c_j \right) \quad (9.39)$$

The values of η_k and ρ_k are given by Equations 9.40 and 9.41.

$$\eta_k = E(\rho_k, C_k - 1) - E(\rho_k, C_k) \quad (9.40)$$

$$\rho_k = \sum_{r:k \in r} \nu_r \prod_{j \in r - \{k\}} (1 - B_k) \quad (9.41)$$

Thus, ρ_k is simply the traffic offered to link k under the approximation procedure. It may be proved that Equations 9.42 and 9.43 hold.

$$\frac{d}{d\nu_r} W(\nu; C) = (1 - L_r) \left(\omega_r - \sum_{k \in r} c_k \right) \quad (9.42)$$

$$\frac{d}{dC_k} W(\nu; C) = c_k \quad (9.43)$$

The definition of Erlang's loss formula can be extended to non-integral values of capacity C by linear interpolation and at integer values of $Csubk$ define the derivative of $W(\nu; C)$ with respect to C_k to be the left derivative.

Equation 9.42 shows that the effect of increasing the offered traffic on route r can be assessed from the following rule of thumb: an additional call offered to route r will be accepted with probability $1 - L_r$; if accepted it will earn ω_r directly, but at a cost c_k for each link $k \in r$. The costs c measure the knock-on effects of accepting a call upon the other routes in the network. From Equation 9.43 it follows that the costs c also have an interpretation as shadow prices, with c_k measuring the sensitivity of the rate of return to the capacity C_k of link

k. Note that Equation 9.39 can be rewritten in the form shown in Equation 9.44 where s_r is given by Equation 9.45.

$$c_k = \eta_k (1 - B_k)^{-1} \sum_{r:k \in r} \lambda_r (c_k + s_r) \tag{9.44}$$

$$s_r = \omega_r - \sum_{j \in r} c_j \tag{9.45}$$

Here s_r is the surplus value of a call on route r.

9.3.3 Applications of implied costs

9.3.3.1 *Network dimensioning*

The Equations 9.42 and 9.43 can be used directly as the basis for an efficient hill climbing algorithm to maximise the expression (9.37). Through analogies with deterministic network flow the shadow price interpretation of Equation 9.43 could be used in algorithms to aid capacity expansion decisions (see Key, 1988). The overall approach also has important economic implications for pricing policies and for the apportionment of revenue between different sections of the network operation.

9.3.3.2 *Decentralised adaptive routeing strategy*

Note that the implied cost, c_k, can be written in the form shown in Equation 9.46 (where s_r is given by Equation 9.47) in terms of observables such as the carried traffics on routes and links.

$$c_k = \rho_k \eta_k \sum_{r:k \in r} \frac{\text{carried traffic on route } r}{\text{carried traffic through link } k} (c_k + s_r) \tag{9.46}$$

$$s_r = \omega_r - \sum_{k \in r} c_k \tag{9.47}$$

Hence, we may construct over time estimates of the implied costs c_k and surplus values s_r. This might be achieved by using, for example, moving average estimators. Suppose that the loss probability, L_r, on route r has also been estimated by similar means. A call offered to route r will generate a net expected revenue of $(1 - L_r)s_r$. Should this quantity be negative for any route then that route should be avoided: more revenue will be lost elsewhere in the network than can be generated by accepting calls on this route. Otherwise, traffic should be shared out between possible routes so as to reflect the net expected revenues $(1 - L_r)ssurr$. Routes with higher values of $(1 - L_r)s_r$ than others should receive an increased share of the traffic. Adjustments to routing patterns made on this basis will need to be gradual since the effect of increasing the traffic to a route will be to push up the loss probability of the route and the implied cost c_k along that route, and hence reduce the net expected revenue $(1 - L_r)s_r$.

9.3.4 Network bounds

This section aims to obtain bounds on the performance of dynamic routeing schemes under a fixed stationary pattern of offered traffic. One method makes minimal assumptions concerning the stochastic structure of the system, and within this rather weak framework the bounds are the best possible. Another method considers various ways to improve these bounds for Poisson offered traffic.

9.3.4.1 *Max-flow bound*

For $(\lambda_{ij} : i < j)$, let $F(\lambda)$ be the maximum attained in the linear programme of Equation 9.48, subject to Equations 9.49 to 9.51.

$$\text{maximise } F(\lambda) = \sum_{i<j} \left(x_{ij} + \sum_{k \neq i,j} x_{ikj} \right) \quad (9.48)$$

$$x_{ij} + \sum_{k \neq i,j} x_{ikj} \leq \lambda_{ij} \quad \forall i<j \quad (9.49)$$

$$x_{ij} + \sum_{k \neq i,j} (x_{ijk} + x_{jik}) \leq C_{ij} \quad \forall i < j \tag{9.50}$$

$$x_{ij} \geq 0, \ x_{ijk} \geq 0, \ x_{ikj} = x_{jki} \quad \forall i,j,k \tag{9.51}$$

The interpretation of this problem is as follows. Regard x_{ij} as the direct flow between nodes i and j along the link (i,j), and x_{ikj} as the flow between nodes i and j through the tandem node k. Regard λ_{ij} as the offered traffic between nodes i and j. Then the linear programme has as its objective function the total flow through the network, and has as constraints the limits imposed by the levels of offered traffic and the capacities of the links.

It may be shown that the bound of expression 9.52 holds for the overall network loss probability, L, where Λ is given by Equation 9.53.

$$L \geq 1 - \frac{F(\lambda)}{\Lambda} \tag{9.52}$$

$$\Lambda = \sum_{i<j} \lambda_{ij} \tag{9.53}$$

In this equation Λ is just the total offered traffic to the network and $F(\lambda)$ is the maximum carried load obtained by solving the above linear programme.

Note also that if the arrival streams are in fact deterministic, then the bound can be attained by a dynamic routeing strategy which routes traffic according to a solution to the linear programme.

9.3.4.2 *Erlang bound*

Suppose now that calls arrive to be connected between nodes i and j ($i,j = 1,2,...,J$) as a Poisson process with rates λ_{ij} and that as the pair $\{i,j\}$ varies it indexes independent Poisson streams. Suppose also that a connected call has a duration which is arbitrarily distributed with unit mean, is independent of earlier arrival times and call durations and of the node pair $\{i,j\}$, and is unknown to the dynamic routeing strategy at the time of the call arrival. Let L be the overall network loss probability. Then it may be shown that a lower bound on L is provided by l, the optimum attained in the linear programme of expression 9.54 subject to expressions 9.55 and 9.56.

$$\text{minimise } l = \sum_{i<j} \frac{\lambda_{ij} b_{ij}}{\sum_{i<j} \lambda_{ij}} \tag{9.54}$$

$$\sum_{i \in S, j \notin S} \lambda_{ij} b_{ij} \geq \left(\sum_{i \in S, j \notin S} \lambda_{ij} \right)$$
$$\times E\left(\sum_{i \in S, j \notin S} \lambda_{ij}, \sum_{i \in S, j \notin S} C_{ij} \right) \quad \forall S \subset \{1, 2, \dots J\} \tag{9.55}$$

$$b_{ij} = b_{ji} \geq 0 \quad \forall i < j \tag{9.56}$$

9.4 Dynamic routeing strategies

Advances in the technology of modern telecommunication systems have led to considerable interest in schemes which can dynamically control the routeing of calls within a network. The purpose of such dynamic routeing schemes is to adjust routeing patterns within the network in accordance with varying and uncertain offered traffics, to make better use of spare capacity in the network resulting from dimensioning upgrades or forecasting errors, and to provide extra flexibility and robustness to respond to failures or overloads.

Two approaches in particular have received considerable attention. In the United States, AT&T (see Ash, 1981) implemented a scheme called Dynamic Non-Hierarchical Routeing (DNHR) which uses traffic forecasts for different times of day in a large scale optimisation procedure to predetermine a routeing pattern. This pattern may be changed hourly, typically in relation to time zone differences.

In Canada, Bell Northern Research has proposed a scheme called Dynamically Controlled Routeing (DCR), based on a central controller which receives information on the current state of all links at intervals of 5–10 seconds. This is used by the controller to determine a routeing pattern which is then distributed back to the nodes.

More recently further dynamic routeing strategies have been proposed such as Dynamic Alternative Routeing (DAR) for the British Telecom national network (see Stacey, 1987). Key, 1990 gives a recent survey of distributed dynamic routeing strategies.

9.4.1 Dynamic Non-Hierarchical Routeing (DNHR)

The DNHR strategy has been in operation throughout the entire AT&T trunk network since 1987 following initial operation within a sub network since 1984. The strategy maintains fixed sequences of alternative routes for each source destination node pair but where the sequences are changed during the day to follow changes in the offered traffic patterns. It is designed to take advantage of traffic non coincidence whereby not all the nodes in the network reach their peak traffic at the same time of day. In international networks, and in trunk networks of geographically large countries, such hourly non coincidence is due to the presence of multiple time zones.

Figure 9.9 shows an example of how this might operate in practice. Traffic between Boston and Miami during the morning peak period may experience congestion on the direct route. A sequence of alternative routes for this time of day might be to attempt overflowing via Chicago, then to try via Phoenix and then finally San Francisco using links which are only lightly congested with direct traffic thanks to the time zone effect which makes the peak periods for traffic between San Francisco and Boston and between San Francisco and Miami, say, occur a few hours later in the day.

In DNHR these patterns are calculated in a network management centre at intervals of months with weekly updates using data on traffics and circuit capacities provided by the nodes. This procedure is integrated within a large scale optimisation procedure for updating the link capacities to cope with the growth in forecasted traffic demands. By taking advantage of idle capacity which can be found in the network due to traffic non coincidence overall network cost savings can be achieved. In a case study of a 24 node network model

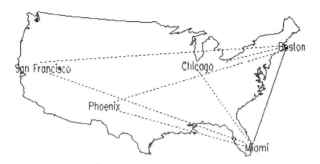

Figure 9.9 DNHR alternative routes

savings can be achieved. In a case study of a 24 node network model it was reported that with DNHR the trunk cost was reduced by about 15% compared with fixed routeing strategies. In international networks even greater savings of the order of 25% have been reported.

9.4.2 Learning automata

Learning automata schemes have been applied to telephone routeing problems (see Narenda, 1983). They continually offer calls across the available routes $r = 1, 2, ..., R$, say, for calls between each source destination node pair according to a probability distribution $(p_r : r = 1, 2, ..., R)$. This probability distribution is updated over time according to the acceptance or rejection of offered calls. A scheme can thereby reward a route on which a call is successful and punish a route on which a call fails.

For example, the learning automation scheme called $L_{R-\in P}$ updates the route selection probabilities as follows. If route i is chosen at time step n and the call is successful then Equations 9.57 and 9.58 are used.

$$p_i(n+1) = p_i(n) + a[1 - p_i(n)] \tag{9.57}$$

$$p_j(n+1) = (1-a)p_j(n) \quad j \neq i \tag{9.58}$$

However, if the call fails, then Equations 9.59 and 9.60 are used.

$$p_i(n+1) = (1-\epsilon)p_i(n) \tag{9.59}$$

$$p_j(n+1) = (1-\epsilon)p_j(n) + \frac{\epsilon p_i(n)}{R-1} \quad j \neq i \tag{9.60}$$

The learning parameter, a, and the penalty parameter, ϵ are two parameters of the scheme such that $0 < \epsilon < a < 1$ and ϵ is small compared to a. The value a is typically small so that the updating step is gradual. If B_r is the probability that a call is blocked on route r then it can be shown that the $L_{R\text{-}\epsilon P}$ scheme tends to approximately equalise blocking probabilities, B_r, while an) equalises the blocking rates $B_r p_r$ over routes $r = 1, 2, ..., R$.

In practice, in a fully connected network it is preferable to first try the direct route between two nodes and then apply the automaton to the choice of two-link overflow routes, rather than to use the automaton to include the single link direct route.

9.4.3 Dynamic Alternative Routeing (DAR)

DAR is a simple, decentralised dynamic routeing strategy that was originally designed for the British Telecom trunk network consisting of between 50 and 60 main switches or nodes which are fully connected. The strategy operates as follows (see Gibbens, 1990). Calls arriving between a source destination node pair first attempt the direct route and are accepted so long as there is at least one free circuit available. If there are no free circuits on the direct route then the call attempts a currently nominated two-link alternative route via a tandem node. The call is attempted on this route with trunk reservation applied against it on both links. If the call is successful then this tandem node remains the currently nominated tandem node for overflow calls. Otherwise, if the call is rejected by this two-link route the call is lost and the tandem node for future overflow calls is re-selected by choosing at random from amongst the set of feasible tandem nodes for that source destination node pair. Note that the tandem node is not re-selected if the call is successfully routed on either the direct route or the two-link alternative route. The term sticky random routeing has been coined to emphasise this property of the strategy.

The DAR strategy attempts to construct a random search algorithm to find and then efficiently utilise spare capacity available within a network under varying conditions of traffic and capacity mismatches.

The simplicity of DAR permits us to extend the Erlang fixed point models of Section 9.3 to model DAR's long run average behaviour. Suppose that n_{ij} is the birth and death process for the utilization of the link (i,j). Then as in Equations 9.25 and 9.26, Equations 9.61 and 9.62 may be obtained where C_{ij} and r_{ij} are the (i,j) link capacities and trunk reservation parameters respectively.

$$q(n_{ij}, n_{ij}+1) = \begin{cases} \lambda_{ij} + \sigma_{ij} & n_{ij} = 0, ..., C_{ij} - r_{ij} - 1 \\ \lambda_{ij} & n_{ij} = C_{ij} - r_{ij}, ..., C_{ij} - 1 \end{cases} \quad (9.61)$$

$$q(n_{ij}, n_{ij}-1) = n_{ij} \qquad n_{ij} = 1, ..., C_{ij} \quad (9.62)$$

Now suppose that $p_k(i,j)$ (where $k \neq \sim i, j$) is the long run proportion of (i,j) overflow traffic offered via node k. Then Equation 9.27 generalises to the asymmetric case of Equation 9.63 where $B_1(i,j)$ and $B_2(i,j)$ are the (i,j) link blocking probabilities for fresh and overflow calls respectively.

$$\sigma_{ij} = \sum_{k \neq i,j} \lambda_{ik} B_1(i,k) p_j(i,k) [1 - B_2(j,k)] \\ + \sum_{k \neq i,j} \lambda_{ij} B_1(j,k) p_i(j,k) [1 - B_2(i,k)] \quad (9.63)$$

Let $L_k(i,j)$ be the blocking on the two-link alternative route via tandem node k given by Equation 9.64.

$$1 - L_k(i,j) = [1 - B_2(i,k)][1 - B_2(k,j)] \quad (9.64)$$

In the case of DAR, it may be shown that the route blocking rates are approximately equalised. Hence expression 9.65, which does not depend on k. Therefore $p_k(i,j)$ may be determined by the relation of Equation 9.66 in terms of $L_k(i,j)$ and hence $B_2(i,j)$ by Equation 9.64.

$$p_k(i,j) L_k(i,j) \tag{9.65}$$

$$\sum_{k \neq i,j} p_k(i,j) = 1 \tag{9.66}$$

The proportions $p_k(i,j)$ may then be used to calculate new estimates for the offered traffics and hence for the link blockings $B_1(i,j)$ and $B_2(i,j)$. Using the method of direct repeated substitution has been found to lead to oscillatory behaviour in some cases. In practice, this may be readily overcome by damping the iteration in the sense that, if solving the fixed point Equation 9.67 for some function f, then one should update x by Equation 9.68 where a ($0 < a \leq 1$) is the damping factor. A value around $a = 0.8$ has been found to be satisfactory in practice.

$$x = f(x) \tag{9.67}$$

$$x_{n+1} = (1-a) x_n + a f(x_n) \tag{9.68}$$

9.4.3.1 *DAR applied to the BT network*

Figure 9.10 shows a typical routeing problem similar to those that were faced in the British Telecom national trunk network. A call requires a route from its trunk exchange (labelled as node 1) to its destination local exchange (node 3). There are two trunk exchanges that connect to node 3 — namely, nodes labelled 2 and 4. This situation, where a local exchange connects to two trunk exchanges, is often repeated throughout the network and is called 'dual-parenting'. First, the call attempts the route (1,2,3) and if a free circuit cannot be found on each of these two links it overflows to the route (1,4,3) via the alternative destination trunk exchange (node 4). If this second route does not have a free circuit available on each link then the call is either lost or further, but necessarily longer, routes are attempted.

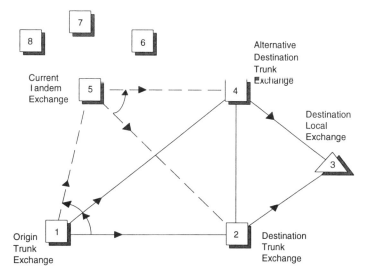

Figure 9.10 DAR in the BT network

A situation where a network may implicitly judge it better to lose the call rather than accept it using routes containing many links is when it is subject to a general overload of traffic. Accepting calls over long routes in these circumstances just adds to the levels of congestion felt by all calls.

The 'sticky random' or Dynamic Alternative Routeing (DAR) strategy handles this situation by remembering a current tandem trunk exchange (labelled node 5) through which to send calls attempting to overflow to longer routes. First, it attempts the route (1,5,2,3) and then overflows to the alternative destination trunk exchange along route (1,5,4,3). For each of these two routes the trunk reservation is used to give priority to the more directly routed calls in preference to the indirect calls using longer routes. If none of these routes is prepared to accept the call then it is at this stage finally rejected from the network. The sticky random strategy then again comes into play to reset the current tandem trunk exchange by choosing at random from amongst a pre-defined set (nodes 5,6,7,8,...). In this way, the sticky random strategy is able to shift the patterns of offered traffic to match the presence of spare capacity that it finds to form these longer paths. It should be pointed out that this strategy applies in parallel, in a decentralised fashion, throughout the network. In an example such as the British Telecom national trunk network there could well be 60 trunk exchanges spread throughout the United Kingdom.

9.4.4 Least busy alternative (LBA)

The least busy alternative routeing strategy is an example of a state dependent routeing strategy defined for fully connected networks which operates as follows. A call is first attempted on the direct route and is accepted so long as at least one free circuit is available. Otherwise, the strategy looks at the occupancies of all the available two-link alternative routes and attempts to route the call along the least busy such route subject to trunk reservation. Specifically, if m_{ik} and m_{kj} give the number of circuits free on the links (i,k) and (k,j) then the alternative chosen is via tandem node k which maximises $\min\{m_{ik}, m_{kj}\}$ over all tandem nodes $k \neq\sim$ i, j. Then trunk reservation with parameter r will mean that the chosen alternative is only used if $\min\{m_{ik}, m_{kj}\} > r$.

It is possible to obtain an Erlang fixed point model for the LBA strategy (see Wong, 1990). We present the model in the simpler case of a symmetric network of N nodes with link capacities C and offered traffics at rate λ. Let n be the birth and death process representing the utilization of a link then the transition rates are as in Equations 9.69 and 9.70.

$$q(i, i+1) = \begin{cases} \lambda + \sigma_i & i = 0, ..., C-r-1 \\ \lambda & i = C-r, ..., C \end{cases} \quad (9.69)$$

$$q(i, i-1) = i \quad i = 1, ..., C \quad (9.70)$$

In this equation σ_i is a state dependent birth rate given in terms of the equilibrium distribution $\{\pi_i : i = 0, ..., C\}$ for n as in Equation 9.71 where y_i is given by Equation 9.72 and φ_i by Equation 9.73.

$$\sigma_i = 2N \left[y_i \sum_{j=0}^{i} \pi_j + \sum_{j=i+1}^{C-1} y_j \pi_j \right] \quad (9.71)$$

$$y_i = \frac{\lambda \pi_C}{N} \sum_{j=0}^{N-1} (\Phi_i)^j (\Phi_{i+1})^{N-1-j} \quad (9.72)$$

$$\Phi_i = 1 - \left(\sum_{j=0}^{i-1} \pi_j \right)^2 \quad (9.73)$$

The loss probability, L is then as in Equation 9.74 where B_1 and B_2 are the link blocking probabilities for fresh and overflow traffics respectively and are given by Equations 9.75 and 9.76.

$$L = B_1 \left[1 - (1 - B_2)^2 \right]^N \quad (9.74)$$

$$B_1 = \pi_C \quad (9.75)$$

$$B_2 = \sum_{i=C-r}^{C} \pi_i \quad (9.76)$$

In the United States, AT&T has recently started implementing a closely related dynamic routeing strategy called Real Time Network Routeing (RTNR) (see Ash, 1991).

9.5 Bibliography

Ash, G.R., Cardwell, R.H. and Murray, R.P. (1981) Design and optimization of networks with dynamic routeing, *Bell Syst Tech J*, **60**, (8), pp. 1787–1820.

Bibliography

Ash, G.R., et al. (1991) Real time network routeing in a dynamic class of service network In *Proc. 13th International Teletraffic Congress*, Copenhagen, North Holland, Amsterdam.

Brockmeyer, E., Halstrom, H.L. and Jensen, A. (1948) *The life and works of A.K. Erlang*, Academy of Technical Sciences, Copenhagen.

Gibbens, R.J. and Kelly, F.P. (1990) Dynamic routeing in fully connected networks, *IMA J. Math Control Inform*, **7**, pp. 77–111.

Hui, Y.J. (1990) *Switching and traffic theory for integrated broadband networks*, Kluwer, Boston.

Kelly, F.P. (1979) *Reversibility and stochastic networks*, Wiley, Chichester.

Kelly, F.P. (1990), Routeing in circuit switched networks: optimization, shadow prices and decentralization, *Adv. in Appl. Probab., 20, pp. 112–144.*

Key, P.B. and Whitehead, M.J. (1988), Cost effective use of networks employing Dynamic Alternative Routeing. *In Proc. 12th International Teletraffic Congress*, Turin, North-Holland, Amsterdam.

Key, P.B. and Cope, G.A. (1990) Distributed dynamic routeing schemes, *IEEE Communications Magazine, Advanced traffic control methods for circuit switched telecommunication networks*, **28**, (10), pp. 54–64.

Narenda, K.S. and Mars, P. (1983) The use of learning algorithms in telephone traffic routeing — A methodology, *Automatica*, **19** (5) pp. 495–502.

Stacey, R.R. and Songhurst, D.J., (1987) Dynamic alternative routeing in the British Telecom trunk network, *Int Switching Symposium*, Phoenix.

Voegtlin, J-C. (1994) The applications of traffic simulation, *Telecommunications*, March

Wong, E.W.M. and Yum, T.S. (1990) Maximum free circuit routeing in circuit switched networks, *Proc. IEEE Infocom '90*, IEEE Computer Society Press, pp. 934–937.

10. Coding

10.1 The need for error control coding

In a digital (discrete) communication system, information is sent as a sequence of digits, which are first converted to an analogue (continuous) form by modulation at the transmitter, and then converted back into digits by de-modulation at the receiver. An ideal communication channel would transmit information without any form of corruption or distortion. Any real channel introduces noise and distortion, however, and these cause the corruption or loss of some digits at the receiver. The system designer can try to reduce the probability of digit errors by appropriate design of the analogue parts of the communication system, but it may be either impossible or too costly to achieve a sufficiently small probability of error in this way. A better solution will often be to use error control coding (ECC).

10.2 Principles of ECC

Error control coding is the controlled addition of redundancy to the transmitted digit stream in such a way that errors introduced in the channel can be detected, and in certain circumstances corrected, in the receiver. It is therefore a form of channel coding, so called because it compensates for imperfections in the channel; the other form of channel coding is transmission (or line coding), which has different objectives such as spectrum shaping of the transmitted signal. ECC is used to lower the probability of error from the input to the output of the communication system. The added redundancy means, however, that extra digits have to be transmitted over the channel, so that either the channel transmission rate must be increased, or the rate of transmission of digits from input to output must be reduced.

The controlled addition of redundancy in ECC contrasts with source coding (data compression) in which redundancy is removed from the source signal.

For a signal (such as speech) with a high degree of intrinsic redundancy it would in principle be possible to perform some error detection and correction at the receiver without adding further redundancy. However this is generally too complex and too dependent on the uncontrolled redundancy of the source signal to be attractive. The functions of error control coding and source coding are therefore usually separated.

The functional blocks used for error control coding are a coder preceding the modulator in the transmitter, and a decoder following the demodulator in the receiver. The decoder may be designed to detect digit errors, or it may be designed to correct them. These functions are known as error detection and error correction respectively.

10.2.1 Types of ECC

There are two main types of ECC: block coding and convolutional coding. In block coding, the input is divided into blocks of k digits. The coder then produces a block of n digits for transmission, and the code is described as 'an (n,k) code'. Each block is coded and decoded entirely separately from all other blocks. In convolutional coding, the coder input and output are continuous streams of digits. The coder outputs n output digits for every k digits input, and the code is described as 'a rate k/n code'.

If the input digits are included unmodified in the coder output the code is described as systematic. The additional digits introduced by the coder are then known as parity or check digits. As well as the conceptual attractiveness of systematic codes, they have the advantage that a range of decoder complexities is made possible. The simplest decoder can simply extract the unmodified input digits from the coded digit stream, ignoring the parity digits. A more sophisticated decoder may use the parity digits for error detection, and a full decoder for error correction. Unsystematic codes also exist, but are less commonly used.

10.2.2 Feedforward and feedback error correction

In feedforward error correction (FEC) the decoder applies error correction to the received codeword, and it may also detect some uncorrectable errors. However, no return path from the receiver to the sender is assumed. Either block or convolutional codes may be used for feedforward error correction.

In feedback error correction, for which only block codes can be used, the receiver only attempts to detect errors, and sends return messages to the sender which cause repeat transmission if any errors are detected in a received block. In the OSI model for packet data networks, this function is carried out within the data link layer, by the return of a positive or negative acknowledgement (ACK or NAK) to the sender on receipt of a data block, a system known as stop-and-wait ARQ. In go-back-N ARQ, receipt of a NAK by the transmitter makes it retransmit the erroneous codeword and the $N-1$ following ones, where N is chosen so that the time taken to send N codewords is less than the round trip delay from transmitter to receiver and back again. This obviates the need for a buffer at the receiver. In selective repeat ARQ, only the codewords for which NAKs have been returned are re-transmitted. Performance analysis (Lin and Costello, 1983) shows that this is the most efficient system, although it requires an adequate buffer in the receiver.

10.2.3 Arithmetic for ECC

In a digital communication system, each transmitted digit is selected from a finite set of M values and is described as an M-ary digit. For example, binary digits (bits) have one of two values, which may be represented as 0 and 1. (The actual values of the physical signal used to transmit the digits, for example +12V and –12V, are irrelevant here). We shall assume that the input message digits use the same value of M as the transmitted digits; if not, the message digits can simply be converted to M-ary before coding.

The analytical design of coders and decoders for M-ary digits requires the use of Galois field arithmetic, denoted GF(M). If M is a prime number (including the important case of binary digits, where

$M = 2$), Galois field arithmetic is equivalent to arithmetic modulo-M. GF(M) arithmetic also exists when M is equal to a power of a prime, so coders using the principles described below can be designed for quaternary (4-valued) and octal (8-valued) digit systems, for example, but in these cases GF(M) arithmetic is not equivalent to modulo-M arithmetic.

In the case of binary systems ($M = 2$), the addition and multiplication operations in GF(2) (i.e. modulo-2) arithmetic are as follows:

$(0 + 0) = (1 + 1) = 0$
$(0 + 1) = (1 + 0) = 1$
$(0 \times 0) = (0 \times 1) = (1 \times 0) = 0$
$(1 \times 1) = 1$.

Clearly, addition in this system is equivalent to the logical exclusive OR (XOR) function, and multiplication is equivalent to the logical AND function. For non-binary systems, multiplication and addition operations can be implemented using either dedicated logic or lookup tables.

Note that in modulo-2 arithmetic, addition and subtraction are equivalent. Some textbooks only discuss binary coders, and therefore treat all subtractions as additions. However for values of M other than 2 addition and subtraction are not equivalent, and it is essential to implement subtractions correctly.

10.2.4 Types of error

If the physical cause of digit errors is such that any digit is as likely to be affected as any other, the errors are described as random. A typical cause of random errors is thermal noise in the received signal. Other types of interference, however, may make it likely that when an error occurs, several symbols in succession will be corrupted; this is known as a burst error. A typical cause of burst errors is interference. Although the true behaviour of the channel may be more complex than either of these simple models, the random error and burst error models are simple, effective, and universally used for describing channel characteristics and error control code performance.

A channel used for transmitting binary digits is known as a binary channel, and if the probability of error is the same for 0s and 1s, the channel is called a Binary Symmetric Channel. The probability of error in binary digits is known as the bit error rate (BER).

10.2.5 Coding gain

Coding gain is a parameter commonly used for evaluating the effectiveness of an error correcting code, and hence for comparing codes. It is defined as the saving in energy per source bit of information for the coded system, relative to an uncoded system delivering the same BER. The effects of both error correction and the increase in transmission rate by a factor of n/k must be included when calculating the coding gain.

10.2.6 Criteria for choosing a code

The primary objective of error control coding will be to achieve a desired end-to-end probability of either uncorrected or undetected digit errors. The choice of code will depend on the error characteristics of the channel (particularly the random and burst error probabilities). The other important factors are likely to be the value of n/k (the increase in transmission rate over the channel), and the implementation complexity and cost of the coder and decoder.

10.3 Block coding

10.3.1 Single parity checks

The simplest block coder appends a single parity digit to each block of k message digits. This produces a systematic $(k + 1,k)$ code known as a single parity code. For binary digits, the parity bit may be either the modulo-2 sum of the message bits or 1 minus that sum. The former case is known as even parity, because the sum of the $k + 1$

bits of the codeword (including the parity bit) is 0, and the latter is known as odd parity.

The decoder forms the sum of the bits of the received codeword. For even parity a sum of 1 means that there has been an error, and a sum of 0 is assumed to mean that the received codeword is correct. (For odd parity 0 indicates an error, and the correct sum is 1.) Note that when an error is detected there is no way to deduce which bit(s) are in error; also, if more than one error occurs, and the number of errors is even, the single parity code will fail to detect them. This code is therefore a single error detecting (SED) code.

10.3.2 Linear block codes

The even parity code is the simplest example of a powerful class of codes called linear block codes. For such codes, the block of k message digits is represented as the k-element row vector d and the n digit codeword produced by the coder is represented by the n-element vector c. The function of the linear block coder is described by the Equation 10.1 where $\mathbf{G}$ is the $(k \times n)$ generator matrix.

$$\mathbf{c} = \mathbf{d\,G} \tag{10.1}$$

The multiplications and additions in this equation are carried out in GF(M) arithmetic (i.e. modulo-2 for binary digits). The codewords generated by the equation are called valid codewords. Since there are 2^k possible datawords, only 2^k of the 2^n possible n-digit words are valid codewords. Systematic linear block codes are produced by a generator matrix of the form shown in Equation 10.2, where $\mathbf{I}_k$ is the $(k \times k)$ unit matrix.

$$\mathbf{G} = [\,\mathbf{I}_k\,|\,\mathbf{P}\,] \tag{10.2}$$

When $\mathbf{G}$ has this form, the codeword c has the form of Equation 10.3, in other words the first k digits of the codeword equal the dataword, and the last $n - k$ digits are parity digits.

$$\mathbf{c} = [\,\mathbf{d}\,|\,\mathbf{d\,P}\,] \tag{10.3}$$

Let r be the received codeword; in the absence of errors $\mathbf{r} = \mathbf{c}$. The decoder performs the operation of Equation 10.4, where $\mathbf{H}$ is the $((n-k) \times n)$ parity check matrix, to produce the $(n-k)$ element syndrome s.

$$\mathbf{s} = \mathbf{r}\,\mathbf{H}^T \qquad (10.4)$$

$\mathbf{H}$ is chosen so that all valid codewords produce a zero syndrome; the syndrome then plays a crucial role in error correction. For the systematic code given above, the optimum form of parity check matrix is simply as in Equation 10.5.

$$\mathbf{H} = [\,\mathbf{P}^T \mid \mathbf{I_{n-k}}\,] \qquad (10.5)$$

For unsystematic codes, construction of the parity check matrix is more difficult.

10.3.3 Distance and code performance

The Hamming distance between two codewords is simply the number of bit positions in which they differ. If the Hamming distance between two codewords $\mathbf{c}_1$ and $\mathbf{c}_2$ is d, and $\mathbf{c}_1$ is transmitted, then d errors would have to occur for codeword $\mathbf{c}_2$ to be received. More generally, if the minimum Hamming distance between codeword $\mathbf{c}_1$ and any other valid codeword is d_M and $\mathbf{c}_1$ is transmitted, then the received codeword will not be a valid codeword if between 1 and $d_M - 1$ errors occur. The decoder could therefore detect up to $d_M - 1$ errors.

When an invalid codeword is received, the distance (number of discrepancies) between it and all the valid codewords can be calculated at the decoder; let the distance between the received codeword and valid codeword $\mathbf{c}_i$ be d_i. If it is found that one of these distances d_j is less than all the others, then assuming random errors it is more likely that the transmitted codeword was $\mathbf{c}_j$ than any other. (This is because for realistic bit error rates the probability that the number of errors will be between 1 and d diminishes rapidly as d increases).

The decoder could therefore be made to output the 'most probably correct' codeword c_j; this is known as Maximum Likelihood or Minimum Distance error correction.

Such error correction is only possible if the number of errors is less than $d/2$, because otherwise the distance to two codewords could be equal, or the distance to a wrong codeword could be less than the distance to the correct one.

In general it is possible to trade off error detection and correction ability; a code which is required to allow correction of n_C errors and detection of a further n_D errors must have a minimum distance given by Equation 10.6.

$$d_M = 2 n_C + n_D + 1 \qquad (10.6)$$

Determining the minimum distance of a code by comparing every pair of codewords would be time consuming for large codeword lengths. The following useful theorem means that only the 2^k valid codewords themselves need to be checked:

'The minimum Hamming distance of a linear block code is equal to the minimum Hamming weight among its non-zero codewords'. (The Hamming weight of a codeword is the number of ones in it).

10.3.4 Hard and soft decision decoding

There are two types of decoding: hard decision and soft decision decoding. In hard decision decoding the demodulated input signal is sliced to produce a digit stream; in the case of binary transmission, slicing is a simple thresholding operation. The decoder performs error detection and correction using this (possibly corrupted) received digit stream.

In soft decision decoding the input to the decoder is the unsliced (analogue) sample stream; since the decoder implementation is usually digital, the sample stream is digitised to an adequate precision before input to the decoder. It has been found that in practice very low resolution digitisation (for example to only 8 or 16 levels) is often adequate. Soft decision decoding is computationally more demanding than hard decision decoding, but is used, particularly with convolutional codes (Section 10.6), to give an extra coding gain, typically of about 2dB.

Soft decision decoding of this kind is hardly ever used with powerful block codes such as cyclic codes (Section 10.5) because of its large processing penalty. However, some codes such as Reed-Solomon codes (Section 10.5.4) can handle erasures (i.e. digits that are known to be in error) as well as erroneous digits, and this can be viewed as a form of soft decision decoding.

10.3.5 Hard decision decoding of linear block codes

The detection and correction of errors is based on analysis of the syndrome. For linear block codes, the syndrome of all valid codewords is zero as in Equation 10.7.

$$\mathbf{s} = \mathbf{c}\,\mathbf{H}^\mathbf{T} = 0 \tag{10.7}$$

If the syndrome of the received codeword is zero, the decoder assumes that it is correct, and the output dataword can be extracted directly from it (very easily, for a systematic code).

A non-zero syndrome is a certain indication of errors. If the transmitted codeword **c** is corrupted by the (modulo-2) addition of the n-bit error pattern **e**, the syndrome becomes as in Equation 10.8, which is a function only of the error pattern.

$$\mathbf{s} = (\mathbf{c}+\mathbf{e})\,\mathbf{H}^\mathbf{T} = \mathbf{e}\,\mathbf{H}^\mathbf{T} \tag{10.8}$$

Hence the syndrome can be used to deduce, and remove, the errors.

There are only $2^{(n-k)} - 1$ different non-zero syndromes, but there are $2^n - 1$ different error patterns, so every syndrome (including zero) may be generated by many different error patterns.

However, as mentioned above, error patterns with small numbers of errors are more likely than those with large numbers of errors, so the most likely cause of a given non-zero syndrome is the corresponding error pattern with the fewest 1s.

This is why the decoder assumes no errors if the syndrome is zero.

One way to implement the decoder is to store in a look-up table (called the Standard Array) the chosen error pattern for each syndrome. The decoder then calculates the syndrome, reads the corresponding error pattern from the table, and subtracts it (using bit-by-bit XOR) from the received codeword. Finally, the output dataword is extracted from the corrected codeword.

10.3.6 Hamming codes

Hamming codes are distance-3 linear block codes, so they can be used for single error correction (SEC) or dual error detection (DED). For binary Hamming codes, the codeword length is given by Equation 10.9, the number of parity bits is r, and the number of message bits is therefore given by Equation 10.10.

$$n = 2^r - 1 \tag{10.9}$$

$$k = n - r \tag{10.10}$$

The first four Hamming codes, for example, are (3,1), (7,4), (15,11), and (31,26) codes.

In Hamming codes, the submatrix **P** in Equations 10.2 and 10.5 is chosen so that its k rows are different from each other, and none of them is all-zero. It is then easy to show that each of the n possible single bit errors generates a different non-zero syndrome, so that any single bit error may be corrected.

10.3.7 2-D parity codes

The single parity code described in Section 10.3.1 is a distance 1, single error detecting (SED) code. A more powerful parity code can be created by arranging the message bits on a rectangular array, and calculating parity bits for each row and each column. Changing any one message bit changes one row parity and one column parity as well, so this is a distance 3-code, which can be used for single error correction or dual error detection.

The SEC decoder works as follows: the sums are formed of each row including its parity bit, and each column including its parity bit. If all the sums are zero, the codeword is assumed to be correct. If only a row sum or only a column sum is 1, the error is assumed to be in the corresponding parity bit. If one row sum and one column sum are 1, the error is assumed to be in the databit in that row and column, which is therefore corrected. If 2 or more row sums, or 2 or more column sums, are 1, a larger number of errors has occurred, which cannot be corrected.

10.3.8 Other codes

The most important linear block codes are the cyclic codes, which are considered in Section 10.5. In addition, many other linear and non-linear block codes have been developed. These include:

1. Hadamard codes. The code words are the rows of a Hadamard matrix, which is a binary $n \times n$ matrix (n even), in which each row differs from any other row in exactly $n/2$ positions; if the matrix elements are denoted +1 and −1, the rows are also orthogonal. Using this matrix and its complement (i.e. the matrix formed by exchanging the +1 and −1 elements), a blocklength n, distance $n/2$ code, with $2n$ code words can be formed.
2. Golay code. The Golay code is a (23,12) triple error correcting (TEC) code.
3. Constant ratio codes. Also known as m-out-of-n codes, have blocklength n, and each codeword has m bits set. These are non-linear codes of primarily historical interest.

10.3.9 Shortened codes

If no convenient code exists with exactly the required value of the dataword length k, a code having a larger value of k, say k' can be shortened. The number of check (parity) bits is unaltered by shortening, but the dataword and codeword lengths are both reduced by the same amount, so the (n', k') code becomes $(n' + k - k', k)$. An example would be the use of a (15,11) Hamming code, shortened to (12,8), for the byte-by-byte transmission of data. Shortening may or may not increase the minimum distance, and hence the error control ability, of the code.

At its simplest, shortening the code consists simply of encoding k' databits, of which k are the required data and $k' - k$ are zero. The main advantage arises when a systematic code is used, because then the first $k' - k$ bits in the codeword can be set to zero and need not be transmitted. In principle, the received codeword can then be lengthened to n' bits again by the reinsertion of $k' - k$ zeros prior to decoding with the (n', k') decoder.

In practice, however, the decoder should be modified, because errors in untransmitted bits are impossible. Consider the example of a shortened Hamming code. The Hamming code can normally correct any single error. However, if on receiving a shortened codeword the decoder calculates a syndrome which would normally imply an error in one of the untransmitted bits, that cannot be the true cause. The next most likely cause is a double error pattern in the transmitted bits. If there were only one double bit error pattern which could produce that syndrome, then the decoder could apply the corresponding double bit correction. If, however, more than one double bit error pattern could have given rise to that syndrome, the decoder cannot guarantee to correct the error. If this is the case, the minimum distance of the shortened code is no greater than that of the unshortened code.

In decoders for shortened cyclic codes, further modifications are desirable to reduce decode time.

10.3.10 Extended codes

An odd-distance code may be extended by adding an overall (even) parity bit; for binary codes this is just the modulo-2 sum of the codebits. This increases the code distance by 1. If the original distance is $2t + 1$, the code is t-error correcting; the extended code can correct up to t errors and detect the case when there are $t + 1$ errors, which is often useful. An example of an extended code is the (24,12) extension of the Golay (23,12) code, which has a convenient rate k/n of exactly 0.5.

10.4 Interleaved and concatenated codes

Interleaving (also known as interlacing) is a technique used to combat burst errors, such as those which occur in fading radio channels. As an illustration, consider ten 15-bit codewords produced by a Hamming (15,11) coder.

Assume that these are stored as the 10 rows of a 10×15 bit array, and that the array is then transmitted column by column. This means that the 10 first bits of the 15 codewords will be transmitted, then the 10 second bits, and so on. The receiver re-groups the received bits into the original pattern of ten 15-bit codewords before decoding.

In this example, if there is a burst error of up to 10 bits duration, no more than one bit in each codeword will be affected. Because the code is single-error-correcting, the receiver can clearly correct these errors, so the use of 10-way interleaving allows correction of 10-bit burst errors.

The extension of this technique to more powerful codes is straightforward, and permits correction of combinations of random and burst errors.

Concatenated coding means the application of one error correcting code to the input data, followed by the application of another code to codewords output by the first coder, and so on. Concatenated coding is valuable when the different codes can combat different types of error (e.g. burst vs. random errors). It is used when the overall efficiency (k/n) of the concatenated code is higher than that of any single code with the same error correcting ability.

10.5 Cyclic codes

Cyclic codes are a very important class of linear block codes, for two reasons: firstly, cyclic codes are available for a wide range of error detecting and correcting requirements, including burst error correc-

10.5.1 The mathematics of cyclic codes

In the theory of cyclic codes, polynomial notation is used rather than the vector notation used earlier. In this notation the k digits of the input data word are treated as the coefficients of a polynomial $d(x)$. For example the 4-binary digit data word (1011) is represented by the polynomial of Equation 10.11. Similarly, the n digit codeword is represented by polynomial $c(x)$.

$$d(x) = 1.x^3 + 0.x^2 + 1.x^1 + 1.x^0 \tag{10.11}$$

The mathematical basis of cyclic codes is the 'algebra of polynomials over GF(M)'; this means that when adding, subtracting, multiplying or dividing polynomials, the arithmetic which has to be carried out on the coefficients is done in GF(M). Binary cyclic codes use polynomials over GF(2) so the coefficient arithmetic is modulo-2. In this case, for example, the product of the polynomials given by equations 10.12 and 10.13 is as in expression 10.14.

$$g(x) = 1.x^1 + 1.x^0 \tag{10.12}$$

$$p(x) = 1.x^2 + 1.x^1 + 1.x^0 \tag{10.13}$$

$$1.x^3 + (1+1).x^2 + (1+1).x^1 + 1.x^0 \tag{10.14}$$

But Equation 10.15 holds, resulting in Equation 10.16.

$$GF(2)(1+1) = 0 \tag{10.15}$$

$$g(x)p(x) = 1.x^3 + 0.x^2 + 0.x^1 + 1.x^0 \tag{10.16}$$

The variable x never needs to be evaluated; it serves only a 'place keeping' role.

An (n,k) cyclic code for M-ary digits is completely defined by a generator polynomial. This generator polynomial must be a factor of $x^x - 1$ and it must be of degree r given by Equation 10.17. In other words it is of the form given by expression 10.18.

$$r = (n - k) \tag{10.17}$$

$$1.x^r + g_{r-1} x^{r-1} + \ldots + g_0 x^0 \tag{10.18}$$

It therefore has $r + 1$ coefficients, the most significant of which is always 1. Published tables are available which give, for each known binary cyclic code, the values of n and k, the coefficients of $g(x)$, and the distance of the code and hence its error detection and correction ability. In such tables it is common practice to represent the binary coefficients of the generator polynomial in actual notation. For example, if the generator of a (7,3) cyclic code (which therefore has 7–3+1=5 coefficients) is written '35' in octal, the binary coefficients are (11101), and the generator polynomial is given by Equation 10.19.

$$g(x) = 1.x^4 + 1.x^3 + 1.x^2 + 0.x^1 + 1.x^0 \tag{10.19}$$

Unsystematic codewords $c(x)$ can be generated from input data words $d(x)$ by calculating Equation 10.20.

$$c(x) = g(x) d(x) \tag{10.20}$$

Of much more interest are systematic cyclic codes, in which the first k bits of the codeword are equal to the data word and the last r bits are parity bits. In polynomial notation Equation 10.21 is obtained, where the parity check polynomial, $p(x)$ is set equal to the remainder after dividing $x^r d(x)$ by $g(x)$.

$$c(x) = x^r d(x) + p(x) \tag{10.21}$$

This remainder operation can be conveniently calculated by feedback shift register circuits described in the next section.

242 Cyclic codes

If the received, possibly corrupted, codeword is $r(x)$, the decoder calculates a syndrome polynomial, $s(x)$, which is used for error detection and correction. $s(x)$ is defined in one of two ways: as the remainder after dividing either $r(x)$ or $x^r r(x)$ by $g(x)$; the second of these two options allows maximum commonality between the coder and decoder, as explained below. In either case a valid codeword produces a zero syndrome.

10.5.2 Implementation of cyclic coders and decoders

Cyclic coders and decoders are implemented using Linear Feedback Shift Registers (LFSR). The general form of a systematic cyclic coder is shown in Figure 10.1, for the code with generator $g(x)$ given by Equation 10.22.

$$g(x) = 1.x^r + g_{r-1}x^{r-1} + \ldots + g_0 x^0 \qquad (10.22)$$

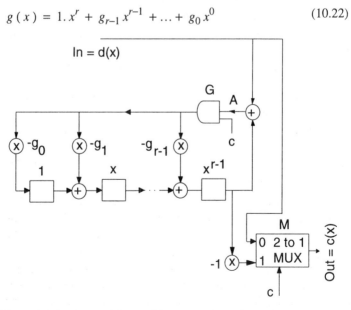

Figure 10.1 General form of a systematic cyclic coder

All multiplications and additions are in GF(M), for example in modulo-2 for binary digits. The rectangular boxes are latches which store the results from multipliers or adders, as appropriate; there are r multipliers and latches (where $r = n - k$ as before). The latches are initially cleared. During the first k cycles of operation control line c is ON and the k digits of the data word are fed through the 2- to-1 multiplexer M to the output; also AND gate G is open, so the adder output A is fed back to the coefficient multipliers. These multiply A by the values $-g_0$ to $-g_{r-1}$. For the last r cycles of operation, control line c is OFF, so gate G is closed and the multiplier inputs and outputs are all zero. The r values stored in the shift register are therefore sequentially output through the 2-to-1 multiplexer.

The function of this circuit is equivalent to the polynomial division $x^r d(x)/g(x)$; the digits produced at A are the quotient of the division, and the contents of the shift register after the kth cycle are the remainder (x^0 term at the left, x^{r-1} term at the right).

For binary coders, all the coefficients are 0 or 1; multiplication by 0 is equivalent to removal of the corresponding path and adder, and multiplication by –1 is equivalent to multiplication by 1 (in modulo-2 arithmetic), which in turn is equivalent to a direct connection to the adder input. Each adder is a modulo-2 adder, or XOR gate. Hence the systematic binary coder for the (7,4) code with generator polynomial given by Equation 10.23 (which would be represented in tables as 1011 in binary, or 13 in octal) is as shown in Figure 10.2.

$$g(x) = x^3 + x + 1 \tag{10.23}$$

The decoder calculates the syndrome of the received (possibly corrupt) codeword, using a linear feedback shift register circuit exactly as in the coder. If the received codeword is represented as $r(x)$, the circuit performs the polynomial division $x^r r(x)/g(x)$, and the contents of the shift register after the nth cycle are the remainder of this calculation, i.e. the required syndrome. Other syndromes are sometimes used; in particular it is possible to produce a circuit to perform the division $r(x)/g(x)$. This is achieved by removing the input from the right-hand end of the LFSR circuit and instead placing an adder at the

244 Cyclic codes

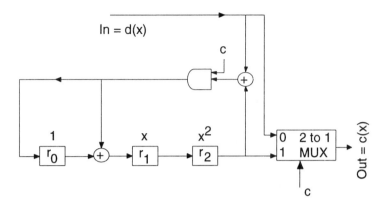

Figure 10.2 A systematic binary cyclic coder

left-hand end, one of whose inputs is the input data word, and the other the output of the $-g_0$ multiplier. This produces a different syndrome, but the decoder operation is the same in either case.

In a decoder which is only designed to detect errors, the syndrome is tested using an r-input OR gate, whose output is 1 (indicating an error) if the syndrome is non-zero. In an error correcting decoder, further processing is required to correct the error. The general form of a cyclic error correcting decoder of the Meggitt or error trapping variety is shown in Figure 10.3. During the first n cycles control line c is ON and the syndrome is calculated as already described; while this is happening the received codeword is stored in a buffer register. During the second n cycles, c is OFF and the received codeword is output from the buffer register. The error pattern detector circuit outputs a 1 (in the binary case) whenever the corresponding bit of the received codeword has been deduced to be in error. This inverts, and hence corrects, the output codeword. Since the code is systematic, its first k digits are the required data word. It can be seen from Figure 10.3 that the signal from the error pattern detector circuit is also fed back to the linear feedback shift register input. This produces an 'updated' syndrome; after the whole codeword has been corrected the remaining syndrome in the shift register should be zero. If it is not, it indicates that more errors have occurred than can be corrected by the

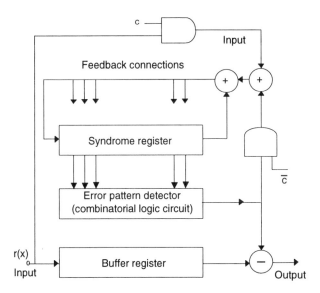

Figure 10.3 General form of a Meggitt or error trapping cyclic error correcting decoder

code. Depending on the system requirements, this condition can be detected and flagged.

The internal details of the error pattern detector are simple for single error correcting codes. In this case, the syndrome corresponding to an error in the first (most significant) bit of the codeword is calculated. The error pattern detector is then simply a combinational logic circuit which outputs a 1 when it detects this pattern in the shift register. For multiple error correcting decoders, as discussed in the following section, the design of the error pattern detector is more complex.

10.5.3 BCH codes

BCH codes are the most extensive and powerful family of error-correcting cyclic codes, and because of their mathematical structure allow decoders to be implemented reasonably easily, even for multiple error correction. Both binary and non-binary BCH codes exist; there are two classes of binary BCH codes: primitive BCH codes, which have a block length $n = 2^m - 1$, and non-primitive BCH codes, where n is a factor of $2^m - 1$. For any positive integers m and $t(t < 2^{m-1})$ there exists a primitive BCH code with the parameters n and r given by Equations 10.24 and 10.25 and minimum distance by Equation 10.26.

$$n = 2^m - 1 \tag{10.24}$$

$$r = n - k \leq m\,t \tag{10.25}$$

$$d_M \geq 2\,t + 1 \tag{10.26}$$

Such a code can therefore be used as a *t*-error-correcting code.

Detailed discussion of both the algebraic basis of BCH codes and the decoder algorithms is described in many textbooks. Primitive BCH codes include cyclic forms of the Hamming SEC codes, and non-primitive BCH codes include a cyclic form of the (23, 12) Golay TEC code.

The two main procedures used in decoding are:

1. Peterson's direct solution, suitable for up to about 6-error-correction.
2. The Berlekamp/Massey algorithm, an iterative algorithm applicable to any BCH code.

Some decoder algorithms have also been developed for specific BCH codes, for example, the Kasami algorithm for the (23, 12) TEC code.

10.5.4 Reed-Solomon codes

Reed-Solomon (RS) codes are an important subclass of non-binary BCH codes. RS codes have a true minimum distance which is the maximum possible for a linear (n,k) code, as in Equation 10.27. They are therefore examples of maximum-distance-separable codes.

Direct Mail Department
Butterworth-Heinemann
FREEPOST
OXFORD
OX2 8BR

UK

you with information on relevant titles as soon as it is available, please fill in the form below and return to us using the FREEPOST facility. Thank you for your help and we look forward to hearing from you.

What title have you purchased? _____
Where was the purchase made? _____
When was the purchase made? _____
Name (Please Print): _____
Job Title: _____
Street: _____
Town: _____
County: _____ Postcode: _____
Country: _____ Telephone: _____
Company Activity: _____
Signature: _____ Date: _____

* Please arrange for me to be kept informed of other books, journals and information services on this and related subjects (* delete if not required). This information is being collected on behalf of Reed International Books Ltd and may be used to supply information about products produced by companies within the Reed International Books group.

(FOR OFFICE USE ONLY)

B UTTERWORTH
H EINEMANN

Butterworth-Heinemann Limited – Registered Office: Michelin House, 81 Fulham Road, London, SW3 6RB. Registered in England 194771.
VAT number GB: 340 242992

$$d = n - k + 1 \tag{10.27}$$

For decoding RS codes, both Peterson's method and the Berlekamp-Massey algorithm can be used. The latter is also known as the FSR synthesis algorithm, because it is equivalent to the generation of the coefficients of a certain LFSR. A further technique for decoding RS codes is called transform decoding, which uses a finite field analogue of the Fourier transform.

A particularly important ability of RS codes (and non-binary BCH codes in general) is their ability to perform error and erasure decoding. This is of value if the receiver can under some circumstances signal the loss of a received digit. If such a code has distance d, it can correct combinations of l errors and s erasures provided that $2l + s < d$.

While RS codes are naturally suited to non-binary digits they can sometimes be used very effectively for binary channels by treating groups of m bits as 2^m-ary digits. It has been shown that such codes outperform binary codes with the same rate and block length at low output error rates. Also, when used in this way RS codes have a natural burst-error correcting ability, because, for example, a burst of up to m bit errors will affect at most 2 'digits'.

10.5.5 Other cyclic codes

10.5.5.1 *Majority logic decodable codes*

These codes are slightly inferior to BCH codes in terms of error correction, but have simple decoder implementations in which the error pattern detector circuit is a combination of XOR and majority logic gates.

10.5.5.2 *Burst error correcting codes*

Fire codes are a widely used class of algebraically constructed burst error correcting cyclic codes, which require a minimum of $3b - 1$ parity bits to correct bursts of up to b bits in length. Other, more efficient, burst error correcting cyclic codes, found by computer search, are also available.

10.5.6 Shortening cyclic codes

Cyclic codes may be shortened; the design of a systematic cyclic coder is unaffected by this, except that only k bits are clocked in rather than k' before the control line is switched to output the parity check bits. In the decoder, the same approach of simply reducing the number of clocks can be used for syndrome polynomial calculation; however, if error correction is to be carried out the error pattern detector must be modified to deal with the missing leading bits, and the fact that there cannot be errors in those missing bits.

10.6 Convolutional codes

A convolutional coder operates on the source data stream using a 'sliding window' and produces a continuous stream of encoded symbols. Many of the concepts related to block codes also apply to convolutional codes; they can for example be systematic or unsystematic, and can be used for error detection or correction. However, there are many good block codes which can be used for error detection with simple encoding and decoding, so when only error detection is required, block codes are almost always used.

By contrast with the analysis and design of block codes, much less mathematical analysis is needed for convolutional codes, and some of the best codes are found by computer search. A rigorous mathematical treatment is however available in Forney (1970). Tables of good convolutional codes for various values of k, n, and L are available in reference books, together with measures of their performance, such as their coding gain at different output BERs.

10.6.1 Convolutional coding

A convolutional coder takes M-ary input digits in groups of k at each time-step and produces groups of n output digits. Since the input and

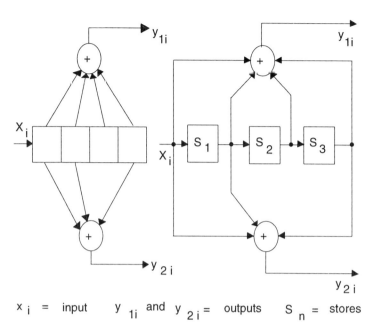

x_i = input y_{1i} and y_{2i} = outputs S_n = stores

Figure 10.4 Two commonly used representations for convolutional coders, for the example of a binary rate 1/2 constraint length 4 code

output data rates are k and n digits per time-step, the code is known as a rate k/n code. The coder contains k parallel shift registers of maximum length $L - 1$, where L is known as the constraint length of the code. Two common pictorial representations of such a coder (with $k = 1$, $n = 2$ and $L = 4$) are shown in Figure 10.4.

The output digits are formed as weighted sums of the input digit(s) and the digits in the shift register(s), with arithmetic in GF(M) as for block codes. In binary, the only possible weights are 0 and 1, so the output bits are formed by selected modulo-2 sums of the input bit(s) and register contents. After the output digits have been formed, the input digit(s) are shifted into the shift register(s).

An alternative representation of a convolutional coder is as a single shift register of length $\leq (L-1)k$ digits. In each time-step the n output digits are calculated and then the k input digits are loaded sequentially into this register. Sometimes Lk is called the constraint length (a conflicting definition with the one above).

The digits in the shift register(s) (s_1 to s_3 in Figure 10.4) define the state of the coder at each time-step; there are therefore $M^{(L-1)k}$ possible coder states. The possible transitions from a given coder state at one time-step to another state at the next time-step are determined by the shift register arrangement. The actual output code digit stream is determined by both the shift register arrangement and the particular summations chosen to form the output digits.

There are M^k patterns of k input digits, so from each state there are M^k possible next states. The behaviour of the coder can therefore be described using a trellis diagram, as shown in Figure 10.5. In this, the possible states of the coder are represented as $M^{(L-1)k}$ nodes in a vertical column, and the state-to-state transitions at each time-step are represented by a rightwards step of one column. From each node at time-step i there are M^k branches to successor states at time-step $i + 1$. In a worthwhile code no two different patterns of input digits give the same transitions.

10.6.2 Viterbi decoding

The error correcting power of a convolutional code arises from the fact that only some of the possible sequences of digits are valid outputs from the coder; these sequences correspond to possible paths through the trellis. The job of the decoder is to find which valid digit sequence is closest to the received digit sequence. This is analogous to the job of a block decoder, but because the input digit sequence is continuous the coder must operate continuously, and must have an acceptably small delay between the arrival of particular input digits and output of the corresponding decoded digits.

The optimum (in the sense of Maximum Likelihood) decoding of convolutionally coded sequences can be carried out using the Viterbi algorithm, which can be applied to both hard-decision and soft-decision decoding. Calculating the most likely digit sequence output by the coder is equivalent to calculating the most likely path followed by

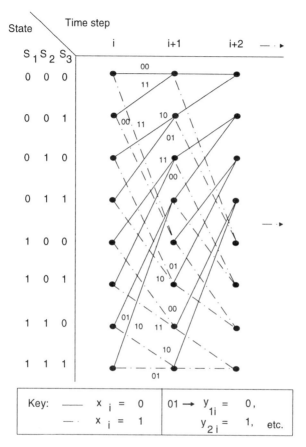

Figure 10.5 The trellis diagram for the coder of Figure 10.4

the coder through the trellis. In a hard-decision decoder (and assuming a binary symmetric channel for simplicity) the most likely coder path is the one which has the smallest number of disagreements with the received bit stream. In a soft-decision decoder the most likely path is the one with the smallest squared difference from the received signal. These measures of discrepancy between the received signal and possible transmitted signals are referred to as path metrics.

It appears at first sight that the number of candidate paths grows exponentially with each time-step. However, at time-step i the decoder only needs to keep track of the single best path so far to each node; any worse path to that node cannot be part of the overall best path through the node. The number of paths to be remembered therefore remains constant and equal to the number of nodes. For each node the decoder must store the best path to that node and the total metric corresponding to that path. By comparing the actual n received digits at time-step $i + 1$ with those corresponding to each possible path on the trellis from step i to step $i + 1$, the decoder calculates the additional metric for each path. From these, it selects the best path to each node at time step $i + 1$, and updates the stored records of the paths and metrics.

In theory the decoder has to wait until the hypothetical end of transmission before it can decide the overall maximum likelihood path from start to finish. Fortunately it can be shown that a fixed finite delay between input and output is all that is necessary to give nearly optimum performance. In particular, it can be shown that if all the reasonably likely paths at time-step i are traced backwards in time, they will converge to the same path within about $5L$ time-steps. The decoder therefore operates with a fixed delay of δ steps, where $\delta \geq 5L$; at time i it selects one of the reasonably likely paths and the bit from that path at time $i - \delta$ is output as the decoded output. The path stores within the decoder can therefore be implemented as δ bit shift registers.

10.6.3 Code performance

A decoding error occurs when the path through the trellis selected by the decoder is not the correct path. The result will almost always be a burst error; it is even possible, for some codes, for the worst error to be infinite in extent. Such codes are clearly unusable. Analysis of the exact statistics of the error bursts (their probability as a function of their duration) is very complex; approximations can however be derived using the minimum free distance of the code. This is obtained by considering all possible error paths, that is, incorrect paths which depart from the correct path at one symbol time and rejoin it later. The minimum free distance is the minimum number of symbol differences between any of these error paths and the correct path. Note, however, that calculation of the true error rates involves analysis of many other interacting effects.

In practice, the user of a convolutional code will normally use the published graphs of output BER as a function of SNR (E_b/n_0). The coding gain is also often quoted, but it is important to remember that this varies with SNR.

10.7 Encryption

10.7.1 Applications of encryption

It is sometimes the case that a sender wishes to send a recipient a message which he wants to keep secret from an eavesdropper. This message, which in its usable form is called plaintext, must be transmitted over a channel to which the eavesdropper is presumed to have access. The process by which the sender and recipient can achieve secrecy is called encryption or encipherment.

In encryption, the plaintext is transformed into a message called the ciphertext in such a way that the recipient can recover the plaintext from the ciphertext, while the eavesdropper cannot. The transformation of plaintext to ciphertext and back is controlled by one or more strings of symbols or digits called keys.

A further use of encryption is for authentication, when it is necessary to check that you are communicating with the correct person and not an impostor.

10.7.2 Principles of encryption

Encryption algorithms have two possible components: substitution, in which plaintext symbols are mapped to different symbols, and transposition, in which the locations of symbols in the ciphertext are altered from the locations of the corresponding symbols in the plaintext. The need for substitution is obvious; many symbols have sufficient significance that an eavesdropper could deduce information from them wherever they were, and perhaps alter them. Transposition is desirable to prevent the eavesdropper deducing information by comparing messages, or corrupting known parts of a message, even if he does not know what they have been altered to.

The most completely secure system is known as the one-time pad. This is a symbol-by-symbol substitution system in which the key is as long as the message; hence no deductions about one part of the ciphertext help the eavesdropper to decipher the rest. The problem is in distribution of the key itself (key management), which is now as big a problem as the original problem of sending the message securely.

Practical encryption systems must have manageable keys, and are based on algorithms for which it is too difficult, rather than theoretically impossible, to decrypt the ciphertext. Various (worst case) assumptions are made in designing and analysing the security of the system; it is normally assumed that the eavesdropper knows the encryption and decryption algorithms, but not the key. When a very high degree of security is required, it is also assumed that the eavesdropper has obtained (by other means) the plaintext corresponding to some of the intercepted ciphertext.

In conventional or symmetric cryptosystems, it is easy to deduce the deciphering key from the enciphering key (they may even be the same), so both must be kept secret. In public key cryptosystems, however, the enciphering key can be made public by the recipient; despite this, it is believed to be computationally infeasible for an eavesdropper to work out the decipherment key. A further technique, public key-distribution (Diffie, 1976) applies the same principles to the secure distribution of keys for conventional cryptosystems.

In stream ciphers, a stream of plaintext symbols is enciphered symbol by symbol. Instead of a true one time pad, the stream of enciphering symbols is generated by a symbol generator under the control of the control of the enciphering key. A simple example for binary data is the use of a pseudo random binary sequence (PRBS) generator, initialised by the chosen key. This produces a stream of 0s and 1s, which are XORed with the databit stream. Clearly, recovery of the original bit stream requires exactly the same operation, so the decipherment key is the same as the encipherment key. Such systems are widely used in communications; their advantages include ease of encryption and decryption, and the fact that single bit errors in the channel cause only single bit errors in the decrypted plaintext. Disadvantages include the effort of initial synchronisation and resynchronisation if synchronisation is lost, and the fact that there is no transposition element in this system.

In block ciphers, the message is divided into n-bit blocks and the blocks are input sequentially to the algorithm. At each stage the key (which is usually constant for the whole message) is also entered, and the encryptor typically uses both substitution and transposition to produce consecutive blocks of ciphertext. In such systems an error in transmission usually results in corruption of the whole received block.

In all these systems, it is essential that the key is changed frequently; the security of an otherwise satisfactory cryptosystem may be compromised if the key is reused.

10.7.3 Specific cryptosystems

Many cryptosystems have been developed, both by government establishments and commercial organisations. The first public standard system was the American National Bureau of Standards Data Encryption Standard (DES), which is a block cipher with a 64-bit blocklength involving both substitution and transposition under the control of a 56-bit key (NBS, 1977), the original proposal was for a 64-bit key and there is debate about whether the 56-bit key is secure enough. Many hardware and software implementations of this standard exist.

The RSA algorithm (Rivest, 1978) is a public-key encryption algorithm in which the recipient publishes an encipherment key N, consisting of the product of two primes, each of order 10^{100}, together with another number E. The sender breaks the plaintext into blocks which can be represented by numbers less than N, then encrypts the blocks using a simple modulo-N arithmetic operation involving E. The recipient recovers the plaintext using further simple modulo-N operations which rely on knowledge of the factors of N. The difficulty of factoring numbers of order 10^{200} means that the eavesdropper cannot work out the two factors of N and cannot decrypt the message.

10.8 Spread spectrum systems

10.8.1 Applications

Spread spectrum systems (Proakis, 1989; Dixon, 1976) are used in digital communications for:

1. Combating interference arising from jamming other users, or self-interference due to multipath effects.
2. Making it difficult to detect the signal, to achieve covert operation.
3. Making it difficult to demodulate the signal, to achieve privacy.

Signals in which spread spectrum techniques are used to make detection difficult are referred to as Low Probability of Intercept (LPI) signals. The use of spread spectrum techniques to allow several users to share a common channel is known as code division multiple access (CDMA) or spread spectrum multiple access (SSMA). In conventional approaches to multiple access (such as FDMA) the channel transfer function may be very poor at some frequencies and good at others. The advantage of using spread spectrum signals for CDMA and for combating multipath is that by using a much wider channel bandwidth, all users get a uniformly acceptable service, rather than a good service for some, and bad for others. For this reason there is currently interest in spread spectrum systems for wireless LANs within buildings.

10.8.2 Direct sequence spread spectrum

In direct sequence spread spectrum (DSSS), the transmitter and receiver contain identical pseudo-random sequence generators producing a pseudo-noise (PN) signal. In the transmitter, the input data stream is XORed with the PN signal before transmission. In the receiver the received signal is XORed with the PN stream to recover the original data stream; this is equivalent to correlation with the known PN sequence. There is an obvious analogy between this process and stream ciphering (Section 10.8) but with the crucial difference that in DSSS the PN sequence is at a much greater clock frequency than the data stream. Each bit of the PN sequence is called a chip, and the clock rate of the PN generator is called the chip rate. In practical systems the chip rate is a large integer multiple L of the databit rate. The bandwidth of the transmitted signal is therefore L times greater than that of the data stream.

The DSSS signal gives LPI (a low probability of intercept) because the total signal power is spread over a wide bandwidth and the signal is noise-like, making it hard to detect. In anti-jamming (AJ) applications, the transmitter introduces an unpredictable element into the modulation of the signal, known also to the receiver but kept secret from opponents, as in stream ciphering. This, together with the wide bandwidth of the transmitted signal, makes jamming more difficult than for conventional signals.

In CDMA applications, the various transmitters which are sharing the channel use different fixed PN sequences which are chosen so that their cross-correlation is low. After a receiver has correlated the received signal with its particular PN pattern, the interference from the other PN sequences is therefore low. The chosen sequences must also have noise-like autocorrelation functions, to help the receiver to synchronise correctly to the partially unknown timing of the transmitter. Some often-used sequences with these properties are called Gold and Kasami sequences (Proakis, 1989).

Direct sequence spread spectrum requires the overall channel (including, where relevant, equalisation in the receiver) to have approximately unity gain, pure delay characteristics over the whole signal bandwidth. This is achievable for local radio systems and transmission lines, but can be much harder to achieve over a wide bandwidth in long distance radio links.

10.8.3 Frequency hopping spread spectrum

In Frequency Hopping Spread Spectrum (FHSS), the available channel bandwidth LW is divided into L slots of bandwidth W. In any signalling interval the signal occupies only one slot or a few ($<< L$) slots.

The spreading of the spectrum arises because the active slot frequency 'hops' around pseudo-randomly. Because of the difficulty of maintaining phase references as the frequency hops, FSK modulation and non-coherent demodulation are normally used in FHSS, rather than PSK and coherent demodulation.

In block hopping FHSS, the input signal is first modulated using a conventional binary or M-ary FSK modulator, whose output is one of M frequencies within a bandwidth W. This block of frequencies is then shifted to somewhere in the full bandwidth LW by mixing it with a local oscillator signal derived from a frequency synthesiser controlled by a PN generator.

The receiver has a matching PN generator, frequency synthesiser and mixer, which shift the frequency block down to baseband again, where it is demodulated by a conventional FSK demodulator. An alternative approach, which has better resistance to some kinds of jamming, but requires a more complex demodulator, is called independent tone hopping. In this, each digit value is combined with the output of the PN generator, to control the frequency synthesiser directly. The resulting frequency separation of the tones corresponding to different digit values at any particular time can be up to the full channel bandwidth.

The frequency hopping rate is usually chosen to be equal to or faster than the symbol rate. If they are equal, it is referred to as slow-hopping, and if the hop rate is faster it is referred to as fast hopping.

Fast hopping is used to combat a follower jammer, which attempts to detect the tone(s) and immediately broadcasts other tones with adjacent frequencies. However, when fast hopping is used, the fact that phase coherence cannot be maintained across hops means that the energy from the successive hops in one symbol must be combined incoherently; this causes a non-coherent combining loss.

FHSS signals are primarily used in AJ and CDMA systems. Usually, FHSS is preferred over DSSS because it has less severe timing requirements and is less sensitive to channel gain and phase fluctuations.

10.9 References

Boyd, C. (1993) Modern data encryption, *Electronics & Communication Engineering Journal*, October.

Burr, A.G. (1993) Block versus trellis: an introduction to coded modulation, *Electronics & Communication Engineering Journal*, August.

Diffie, W. and Hellman, M.E. (1976) New direction in cryptography, *IEEE Trans. Inform. Theory*, pp. 644-654.

Dixon, R.C. (1976) *Spread Spectrum Systems*, Wiley, New York.

Forney, G.D. (1970) Convolutional codes 1: Algebraic structure, *IEEE Trans. Info. Theory*, (IT-16) pp. 720-738.

Lin, S. and Costello, D.J. (1983) *Error control coding: fundamentals and applications*, Prentice-Hall, Englewood Cliffs, NJ.

Michelson, A.M. and Levesque, A.H. (1985) *Error-control techniques for digital communication*, Wiley-Interscience, New York.

NBS (1977) National Bureau of Standards. *Data encryption standard*, Federal Information Processing Standard (FIPS) Publication No. 46.

Peterson, W.W. and Weldon, E.J. (1972) *Error-correcting codes*, 2nd ed., MIT Press, Cambridge, MA.

Proakis, J.G. (1989) *Digital communications*, 2nd ed., McGraw-Hill, New York.

Rivest, R.L., Shamir, A. and Adelman, L. (1978) A method for obtaining signatures and public-key cryptosystems, *Commun. ACM*, pp. 120-126.

Tsui, T.S.D. and Clarkson, T.G. (1994) Spread-spectrum communication techniques, *Electronics & Communication Engineering Journal*, February.

Princen, M. (1991) Encryption with DES and its implementation in an ASIC, *Electronic Engineering*, January.

Ventham, J. (1991) Security in defence, *Communications International*, May.

11. Signals and noise

11.1 Definition of a signal

The definition of a signal may take a variety of forms:

1. The conveying of information through a medium.
2. The physical embodiment of a message.
3. A media manifestation conveying information or direction from one end of a transmission medium to another.

Two further definitions associated with a signal are :

1. Signal frequency, the frequency of the carrier wave upon which the signal information has been impressed.
2. Signalling, the use of a signal to convey coded direction or instructions to a person or piece of apparatus at a distance, such a signal normally being associated with the establishment, servicing or breaking of a connection.

Signals that follow the instantaneous variation of the original information energy are defined as analogue. A digital signal is in the form of a pre-determined code of pulses or variations, which represent symbols taken from a selected set of symbols.

A direct current signal is when the flow of current is only in one direction. However, the strength of the current may be varied. A direct current can be produced from an energy source such as a dry battery.

A steady direct current flowing in a circuit cannot convey information. The inclusion of a simple on-off switch enables the current to be regulated into a series of pulses. These pulses represent symbols which have an agreed meaning at the transmitting and receiving ends of the link. A series of the symbols can be concatenated to form a message.

If the connections to the direct current source are reversed, then the direction of the current is also reversed. The current direction can therefore be considered to be positive or negative according to the way it is flowing around the circuit, and this is known as the polarity of the circuit.

There are two principal methods for creating signals with direct current sources. Firstly the information may be carried by the alternate presence and absence of current or, secondly, by the switching of the direct current sources between two distinct values. The feature common to both of these methods is the variation in the amplitude of the current.

A typical use of direct current signal, using the alternate presence and absence of a pulse, is between a telephone and the local exchange.

The state of the signal indicates on-hook, off-hook, dial pulses, or the status of the connection. On-hook is indicated by an open circuit and no current flow. Off-hook is indicated by a closed circuit and continuous current flow. Dial pulses consist of current flow interrupted at a specified rate.

There are various difficulties associated with direct current signalling :

1. Circuits that have long transmission lines are subject to attenuation and distortion, although these can be rectified using regeneration and amplification.
2. Connecting wires are needed for the whole of a telecommunications circuit.

A rotating alternator or an electronic oscillator causes the current to reverse direction at regular intervals to create a particular repeating pattern or waveform. This is known as an alternating current. The advantages of alternating current signals are :

1. The strength and amplitude can easily be altered allowing transmission over long lines.
2. Connecting wires are not necessary for the whole of a telecommunications circuit.

Table 11.1 Example of a dibit operation

Bibits	Phase
00	+90
01	0
10	+180
11	+270

11.1.1 Bits and bauds

The bit rate is defined as the number of binary digits that are transmitted through a transmission medium per second. The baud is the unit of modulation rate. When a modulation method is used that has binary values then the line rate for the modulation directly corresponds with the transmission of bits, and the bit rate and baud rate are equal. When a multi-level modulation method is used each modulation can represent more than one bit. If one modulation equals two bits, each modulation represents a dibit. A tribit is when each modulation represents three bits. In the latter case the bit rate is three times the modulation rate.

An example of the dibit operation is found in ITU-T (formerly CCITT) V.22 recommendations which uses four phase changes for modulation, as in Table 11.1. The baud rate is 600 whilst the bit rate is 1200 per second.

ITU-T V.29 uses 4 bits per baud which are carried using 8 different phase changes, 45 degree apart, and two different amplitude changes.

11.2 Waveform and frequency

There are a wide variety of waveforms that are possible with alternating currents. One of the simplest to produce, a sinusoidal waveform, shown in Figure 11.1, is created by rotating a loop of wire in a uniform magnetic field.

Signals and noise 263

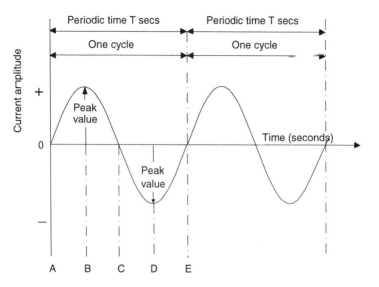

Figure 11.1 Sinusoidal alternating current waveform

In Figure 11.1 between points A and B the current increases from zero to a peak value in a positive direction. Between points B and C the current gradually reduces to zero. Then between points C and D the current moves to a peak value in the opposite or negative direction. Finally between points D and E the current gradually returns to zero again. The whole sequence from point A to point E represents one complete rotation of the wire loop in the magnetic field and is called one cycle of an alternating current waveform.

The periodic time, which is measured in seconds, is the time required for one complete cycle of the alternating current waveform to be produced.

The frequency of the waveform is the number of complete cycles that occur in one second and is therefore the reciprocal of the periodic time. Frequency is measured in Hertz (Hz).

264 Waveform and frequency

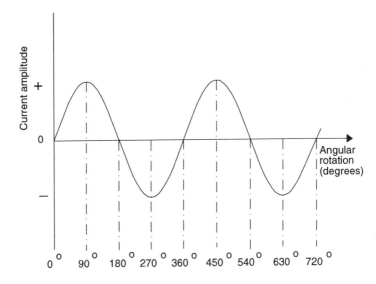

Figure 11.2 Angular rotation

The amplitude of the waveform represents the strength of the current at any instance, with the direction of the current, positive or negative, being referred to as the polarity.

The rotation of the wire loop within the magnetic field can be expressed in terms of rotation. The wire loop for each cycle will move through 360 degrees. Figure 11.2 shows the sinusoidal waveform plotted against angular rotation.

At the same point in each cycle the amplitude of the waveform has the same value. The identification of a particular point in the cycle as the degree of rotation is called the phase of the alternating current waveform.

Phase difference describes two waveforms that are identical except for their phase. Figure 11.3 illustrates two sinusoidal waveforms that have a phase difference.

Amplitude, frequency and phase represent the characteristics of a waveform, that may by varied so that symbols may be transmitted along the media, thus allowing a complete message to be sent.

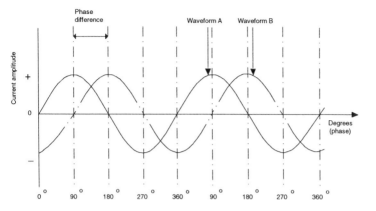

Figure 11.3 Phase difference

The energy of an alternating current waveform will travel along a transmission medium at a particular velocity, therefore a certain distance will be travelled in one cycle. The alternating current waveform repeats complete cycles over equal distances. The distance representing each cycle is called the wavelength and is conventionally measured in metres.

Equations 11.1 to 11.3 represent the relationships between velocity, wavelength, time and frequency.

$$\text{Velocity} = \frac{\text{Wavelength}}{\text{Time}} \tag{11.1}$$

$$\text{Frequency} = \frac{1}{\text{Periodic time}} \tag{11.2}$$

$$\text{Velocity} = \text{Wavelength} \times \text{Frequency} \tag{11.3}$$

11.2.1 Waveshapes

All component waveforms are made of a sinusoidal waveform which has a certain frequency, called the fundamental frequency and a

266 Digital signals

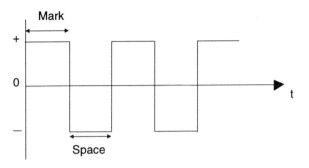

Figure 11.4 Square waveform

number of other sinusoidal waveforms having frequencies that are direct multiples of the fundamental frequency. These direct multiples are harmonics of the fundamental frequencies.

For a complex waveform having a fundamental frequency, f, the harmonics with multiple values of f (e.g. $2f$, $3f$, $4f$, etc.) may also be present. The square waveform is made of a fundamental frequency and all the odd harmonics rising to infinity, i.e. $3f$, $5f$, $7f$, $9f$, and so on. Figure 11.4 shows the square waveform with the associated terminology of space and mark.

The amplitudes of the various harmonics are added to give a resultant waveshape that approximates to a square wave. The addition of the fundamental with the 3rd and 5th harmonics is illustrated in Figure 11.5. A more exact square wave would be created by the addition of further odd harmonics.

A second common waveshape is the saw tooth waveform which contains a fundamental frequency and all the odd and even harmonics to infinity.

11.3 Digital signals

There are three ways in which a digital signal may be transmitted to a line :

1. A single (unipolar) current.

Signals and noise 267

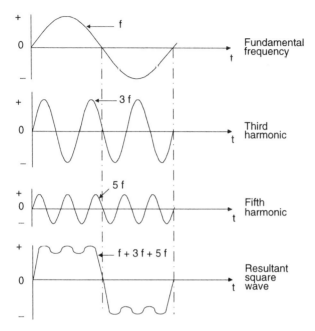

Figure 11.5 Amplitude addition

2. The presence or absence of a voltage.
3. Two voltages (bipolar) one positive and one negative with respect to earth.

The use of digital transmission instead of analogue gives the following advantages:

1. A better signal to noise ratio. Noise and interference are accumulative with distance on an analogue system, but not on a digital system.
2. Improved signalling capability.
3. Time division multiplexing, used with digital signals, is simpler than frequency division multiplexing, which is used with analogue signals.

4. Digital switching is more straightforward to implement.
5. In digital systems different kinds of signals, e.g. data, telegraph, telephone speech and television, have the same representation on the medium and therefore can be treated in a similar manner during transmission and switching.

11.4 Examples of signals

11.4.1 Voice and music

The sounds produced by a human voice are variations of air pressure above and below normal pressure. Such sounds are alternating in nature and have a complex waveform that is different for each voice. The voice contains fundamental frequencies and associated harmonics. Therefore the sound waves produced by human voices contain a range of frequencies which is known as bandwidth.

Voice frequencies lie within the band 100Hz to 10000Hz. The pitch of the voice is determined by the fundamental frequency of the vocal chords and is between 200Hz and 1000Hz for women and between 100Hz and 500Hz for men. The tonal quality and the individuality are determined by the higher frequencies that are produced.

The power content of speech is small, an average being between 10 and 20 microwatts. The distribution of the power is not even, with most of it being contained in the region of 500Hz for men and 800Hz for women.

The notes produced by musical instruments occupy a much larger frequency band than occupied by speech. Some instruments, such as the drum, have a fundamental frequency of 50Hz or less, while others, such as the violin, can produce a note which has a harmonic content in excess of 15000Hz. An orchestra may generate a peak power in the region of 90 to 100 watts.

In practice not all frequencies produced are transmitted to the receiver. There are two principal reasons for this. Firstly, it is more economic to use devices in circuits that have a limited bandwidth and, secondly, on long distance routes a number of circuits are transmitted over a single telecommunications link and this provides a further limitation on bandwidth.

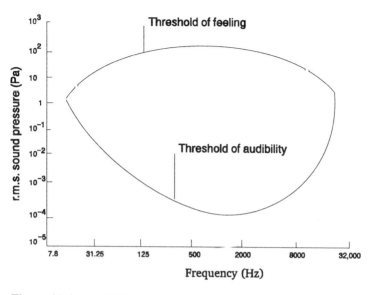

Figure 11.6 Audibility

At the receiving end of the speech or music the sound waves impact on the ear and cause the ear drum to vibrate. The ear can only hear sounds that lie within certain limits. The minimum sound level that can be detected by the ear is known as the threshold of audibility, with the sound level that produces a feeling of discomfort being known as the threshold of feeling.

Figure 11.6 illustrates the thresholds of feeling and audibility by plotting sound pressure against frequency. It can be seen that the most sensitive region is between 1000Hz and 2000Hz although response is capable with the 30Hz to 16500Hz range.

11.4.2 Telephone and telegraph

By international agreement the audio frequency for a commercial quality speech circuit routed over a multi-channel telephone system is restricted from 300Hz to 3400Hz. The suppression of all frequencies above 3400Hz reduces the quality of sound but does not affect the intelligibility.

A telegraphy system is one that passes messages by means of a pre-agreed signalling system. Typical codes that have frequently been used are Morse and Murray, the latter also being known as the International Alphabet 2 (IA2).

The characters in Morse code are represented by 'dot' and 'dash' signals. A dot is differentiated from a dash by differing time periods. Spacings between, signals, words and letters are distinguished by different time periods. Morse code is not adopted for use with automatic receiving equipment as the number of signal elements that comprise a symbol varies, as does the actual length of the signal element. The bandwidth required for Morse code lies in the range of 100Hz to 1000Hz.

The teleprinter system uses the Murray code which has signal elements that are of equal length, and all characters have exactly the same number of signal elements. Each character is represented by a negative potential and a space by a positive potential.

Teleprinters normally operate at a speed of 50 baud. The maximum periodic time for one cycle of an alternating waveform is 40ms, giving a fundamental frequency of 25Hz.

11.4.3 Radio and television

The bandwidth for speech on a circuit operated over a high frequency radio link is 25Hz to 3000Hz. Land line circuits that connect studio and transmission sites have a bandwidth of 30Hz to 10000Hz, which allows very high quality sound. Long and medium wave broadcasting cannot use such a wide bandwidth because the high demand from other broadcasting stations.

By international agreement medium waveband broadcasting stations in Europe are spaced at approximately 9000Hz apart. The receiver is therefore confined to about 4500Hz so as to avoid interference.

The bandwidth for a television picture signal depends on a number of factors such as: the number of lines that make up the picture; the number of fields transmitted per second; and the duration of the synchronising pulses. In general the normal bandwidth is 5.5MHz. The bandwidth of the audio signal is 20Hz. A colour television picture consists of a brightness (luminance) component which corresponds to the monochrome signal plus colour (chrominance) information which is transmitted as amplitude modulation side bands of two colour sub-carriers which are of the same frequency approximately 4.434MHz but 90 degrees apart. No extra bandwidth is needed to accommodate the colour information.

11.4.4 Radar

When a radio wave strikes an object some of the energy in the signal is reflected back to the transmitter so that the object may be detected. If the transmitted radio wave is in the form of pulses, then by measuring the item delay between the transmitted and the received pulse (the echo) the distance of the object can be calculated. Highly directional transmitting aerials also allow the bearing of the distant object relative to transmitter to be calculated.

Radar is different from other transmission methods discussed, since the transmission does not carry actual information. A typical transmitter produces pulses of radio waves in the frequency range of 150MHz and 30000MHz. The duration of the pulses will generally lie between 0.25μs and 50μs. The time interval between the pulses depends on the maximum distance at which the radar system is effective.

11.5 Classification of signals

11.5.1 Energy and power signals

A voltage or current represents a signal in an electrical system. In the time interval t_1 to t_2 the energy dissipated by a voltage in a resistance is given by Equation 11.4 and the current by Equation 11.5.

$$E = \int_{t_1}^{t_2} \frac{V^2(t)}{R} dt \tag{11.4}$$

272 Classification of signals

$$I = \int_{t_1}^{t_2} R i^2(t) \, dt \tag{11.5}$$

A conventional way of referring to the energy is by considering the case of one-ohm resistance. Equations 11.4 and 11.5 then take the same form. A description may then be given of the energy associated with any signal $x(t)$ in dimensionless form, as in Equation 11.6.

$$E = \int_{t_1}^{t_2} x^2(t) \, dt \tag{11.6}$$

An energy signal is one for which Equation 11.6 is finite. The condition that must be satisfied is given in expression 11.7.

$$\int_{-\infty}^{\infty} x^2(t) \, dt < \infty \tag{11.7}$$

There are many signals that do not satisfy expression 11.7. Periodic signals are the principal group although there are also some aperiodic signals that are included in this group. If this expression is not satisfied, then it is more appropriate to consider the average power of the signal. If a one-ohm basis is considered then the average power, from Equation 11.6, is given by Equation 11.8.

$$P = \frac{1}{t_2 - t_1} \int_{t_1}^{t_2} x^2(t) \, dt \tag{11.8}$$

A power signal satisfies the condition given in Equation 11.9.

$$\varphi < \lim_{T \to \infty} \frac{1}{2T} \int_{-T}^{T} x^2(t) \, dt \tag{11.9}$$

Therefore an energy signal has zero average power and a power signal has infinite energy. Thus this type of signal classification is mutually exclusive. However, there are some signals that fit into neither classification.

11.5.2 Periodic and aperiodic signals

A periodic signal is one that repeats the sequence of values exactly after a fixed length of time, known as the period. In mathematical terms a signal $x(t)$ is periodic if there is a number T such that for all t Equation 11.10 holds.

$$x(t) = x(t + T) \tag{11.10}$$

The smallest positive number T that satisfies Equation 11.10 is the period and it defines the duration of one complete cycle. The fundamental frequency of a periodic signal is given by Equation 11.11.

$$f = \frac{1}{T} \tag{11.11}$$

It is important to distinguish between the real signal and the quantitative representation, which is necessarily an approximation. The amount of error in the approximation depends on the complexity of the signal, with simple waveforms, such as the sinusoid, having less error than complex waveforms.

A non-periodic or aperiodic signal is one for which no value of T satisfies Equation 11.11.

In principle this includes all actual signals since they must start and stop at finite times. However, aperiodic signals can be presented quantitatively in terms of periodic signals.

Examples of periodic signals include the sinusoidal signals and periodically repeated non-sinusoidal signals, such as the rectangular pulse sequences used in radar.

Non-periodic signals include speech waveforms and random signals arising from unpredictable disturbances of all kinds. In some cases it is possible to write explicit mathematical expressions for non-periodic signals and in other cases it is not.

In addition to periodic and non-periodic signals are those signals that are the sum of two or more periodic signals having different periods. T will not be satisfied in Equation 11.10, but the signal does have many properties associated with periodic signals and cannot be represented by a finite number of periodic signals.

11.5.3 Random and deterministic signals

Signals may be classified as random or deterministic, the deciding criterion being the predictability of a signal before it is generated. Random or stochastic signals are those which have variations in magnitude that occur in an unpredictable manner. Interference and noise constitute random features of signals.

If the possible future values of a signal can be predicted from the study of previous signals then no suitable quantitative expression can be derived.

Under some circumstances, following the observation of signals, a mathematical expression can be constructed to describe a waveform, but there are some factors associated with parameters such as phase that are unpredictable in the first instance.

A deterministic signal is one about which there is no uncertainty before it occurs and in almost all cases an explicit mathematical expression can be written for it.

11.6 Signal representation

11.6.1 Time and frequency domains

A signal $x(t)$ may be represented in terms of sinusoids having frequencies that are multiples of the fundamental frequency $\frac{1}{T}$. Coefficients are added to give the magnitude and phase of the sinusoids and this combination represents the frequency domain description of the signal.

The explicit time function $x(t)$ is the time domain description of the signal.

11.6.2 Complex representation

Signals may be represented mathematically in different ways. The basis for many waveforms is the sinusoidal form and thus sine and cosine representation is a convenient mathematical method of describing signals.

A second method, often preferred for mathematical convenience, is the representation of a pair of real value signals in complex mathematics.

A complex signal consists of a real signal and an imaginary signal, which may be visualised as two voltages across two resistors. Complex valued signals are processed just as real valued signals, except the rules of complex arithmetic are used. Conventionally j is taken to represent the imaginary part.

The relationship between the sine and cosine representation and the complex representation is given in Equation 11.12, where ω_0 is given by Equation 11.13.

$$e^{\pm j n \omega_0 t} = \cos n \omega_0 t \pm j \sin n \omega_0 t \tag{11.12}$$

$$\omega_0 = \frac{2\pi}{T} \tag{11.13}$$

11.6.3 Fourier series representation

The use of real and complex sinusoids to represent signals are called Fourier methods, after the mathematician who first investigated these techniques.

In the case of signals the Fourier representation has direct physical interpretation through measured quantities. Fourier analysis of signals also lends itself to automatic calculation on a computer based system.

Various waveforms have Fourier series expressions. The series for a general periodic wave that has an arbitrary period T is given in Equation 11.14, the coefficients being given by Equations 11.15 and 11.16.

276 Signal representation

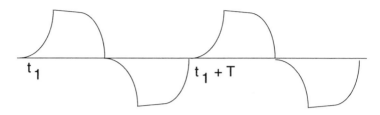

Figure 11.7 Waveform with the Fourier expression given by Expression 11.7

$$\frac{a_0}{2} + \sum_{n=1}^{\infty} a_n \cos \frac{2\pi n t}{T} + b_n \sin \frac{2\pi n t}{T} \qquad (11.14)$$

$$a_n = \frac{2}{T} \int_{t_1}^{t_1+T} f(t) \cos \frac{2\pi n t}{T} dt \qquad (11.15)$$

$$b_n = \frac{2}{T} \int_{t_1}^{t_1+T} f(t) \sin \frac{2\pi n t}{T} dt \qquad (11.16)$$

The waveform depicted in Figure 11.7 has the Fourier series given by expression 11.17 with the coefficient given in Equation 11.18.

$$\sum_{n=-\infty}^{\infty} a_n \varepsilon^{j(2\pi n/T)t} \qquad (11.17)$$

$$a_n = \frac{1}{T} \int_{t_1}^{t_1+T} f(t) \varepsilon^{-j(2\pi n/T)t} \qquad (11.18)$$

There are two varieties of square waveform that can be described by Fourier series, one being odd and the other even. Figure 11.8 illustrates the two waveforms with expression 11.19 giving the Fourier representation of the odd square wave and expression 11.20 giving the representation for the even square wave.

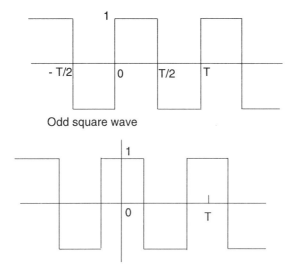

Figure 11.8 Odd and even square waveforms

$$\frac{4}{\pi} \sum_{n=1}^{\infty} \frac{1}{2n-1} \sin \frac{2\pi(2n-1)t}{T} \tag{11.19}$$

$$\frac{4}{\pi} \sum_{n=1}^{\infty} \frac{(-1)^{n+1}}{2n-1} \cos \frac{2\pi(2n-1)t}{T} \tag{11.20}$$

A regular pulse train, illustrated in Figure 11.9, has the Fourier expression given in 11.21.

$$\frac{2t_0}{T} = \frac{4t_0}{T} \sum_{n=1}^{\infty} \frac{\sin \frac{2\pi n t_0}{T}}{\frac{2\pi n t_0}{T}} \cos \frac{2\pi n t}{T} \tag{11.21}$$

278 Signal representation

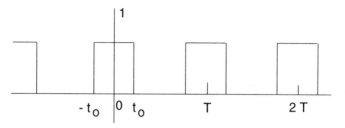

Figure 11.9 Rectangular pulse train with the Fourier expression given by Equation 11.21

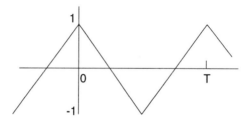

Figure 11.10 Triangular waveform with the Fourier expression given by Equation 11.22

The triangular waveform is illustrated in Figure 11.10 and the Fourier expression is given in Equation 11.22.

$$\frac{8}{\pi^2} \sum_{n=1}^{\infty} \frac{1}{(2n-1)^2} \cos \frac{2\pi(2n-1)t}{T} \tag{11.22}$$

Figure 11.11 illustrates the sawtooth waveform and Equation 11.23 gives the expression corresponding to that figure.

$$\frac{2}{\pi} \sum_{n=1}^{\infty} \frac{(-1)^{n+1}}{n} \sin \frac{2\pi nt}{T} \tag{11.23}$$

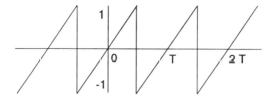

Figure 11.11 Sawtooth waveform with the Fourier expression given by Equation 11.21

11.7 Distortion of signals

11.7.1 Types of distortion

Distortion of a signal may be expressed in a variety of ways:

1. Distortion has occurred if the output signal from a channel is not the exact replica of the input to that channel.
2. Distortion is the difference in time between the output signal of a receiver and the input signal causing that output.
3. Distortion occurs if there is a change in waveform between the input and output terminals.

In general distortion arises when the output waveform contains frequency components that are not present in the original signal, or where complex signals are involved when the phase relationships between the various components of the signal may be altered. The relative amplitudes of the harmonic components may also be altered.

The relationship between the input and output voltage is known as the transfer characteristic. The gain therefore varies with the instantaneous magnitude of the input signal and non-linear distortion occurs.

11.7.1.1 *Attenuation distortion*

If all frequencies involved in the transmission were subject to the same gain and the same loss then attenuation distortion would not occur. However, in all transmission media some frequencies are attenuated more than others, therefore attenuation distortion results from imperfect amplitude frequency response.

Attenuation distortion across the voice channel is measured against an ITU-T reference frequency of 800Hz. In North America 1000Hz is commonly used as the reference frequency. If a signal at 10dBm is placed on the input of a channel, then assuming no gain or loss the output would be 10dBm at 800Hz, whereas the output at 2500Hz could be 11.9dBm and at 1100Hz it could be 9dBm. The attenuation distortion for a voice channel is illustrated in Figure 11.12.

11.7.1.2 *Delay distortion*

The velocity with which a signal travels through a medium is a function of the velocity of propagation, which varies between mediums. The velocity tends to vary with frequency, with an increase towards the band centre and a decrease towards band edge. The finite time the signal takes to pass through the voice channel of the transmission link is called the delay. At a reference frequency the delay a signal experiences is referred to as the absolute delay.

11.7.1.3 *Phase distortion*

With the propagation time being different for different frequencies, the wave front of one frequency will arrive before the wave front of another frequency, thus the phase has been shifted or distorted. The result is that the relative phases of the harmonic components of a signal are not maintained.

11.7.1.4 *Harmonic distortion*

With non-linear distortion the application of a sinusoidal input voltage results in a periodic output waveform that is non-sinusoidal. Fourier analysis shows that spurious harmonics are present, the result being known as harmonic distortion. Total harmonic distortion is measured as the root of the sum of the squares of the r.m.s. voltages

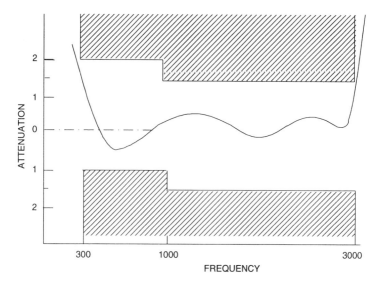

Figure 11.12 Attenuation distortion voice channel

of the individual harmonics divided by the r.m.s. of the total output signal, V, where V_{H2}, V_{H3} are the r.m.s. values of the harmonic components. Equation 11.24 gives the total harmonic distortion.

$$D = \frac{(V^2_{H2} + V^2_{H3} + V^2_{H4} + ... + V^2_{Hn})^{\frac{1}{2}}}{V} \; 100 \; \% \quad (11.24)$$

11.7.1.5 *Intermodulation distortion*

Intermodulation distortion is a form of non-linear distortion whereby the amplitude of a signal of one frequency is affected by the amplitude of a simultaneously applied signal of a lower frequency. Combination frequencies are produced which have values equal to the sum and difference of the two applied frequencies. The non-linear transfer characteristics reduce the amplitude of the higher frequency signals at times when the lower frequency signal is near to maximum and minimum voltages (Figure 11.13).

282 Types of noise

Figure 11.13 Illustration of intermodulation distortion

11.7.2 Distortion of digital signals

11.7.2.1 *Attenuation distortion*

The attenuation of signals in the voice frequency range increases with an increase in frequency and so the various harmonics contained in a digital waveform will be attenuated to a greater extent than the fundamental frequency. The greater attenuation on the higher order harmonics means the rectangular waveshape will gradually be lost. The effect is accentuated by the length of the line and so pulses will be more rounded.

11.7.2.2 *Synchronisation distortion*

Synchronisation is the principle whereby the receiver will at all times sample the incoming bit at the correct instance in time. Distortion can occur if the sampling drifts from the correct sample period. The two methods of synchronisation, isochronous and anisochronous, the latter often being referred to as start-stop, can both suffer from distortion.

11.8 Types of noise

11.8.1 Thermal or Johnson noise

Thermal or Johnson noise, first documented by J.B. Johnson of Bell Laboratories in 1928, is often called white noise. It is of a gaussian nature which means it is completely random. Any circuit operating at a temperature above absolute zero (-273°C) will display thermal

Signals and noise 283

noise. This random noise is caused by the random motion of discrete electrons in the conductive path.

The work of Johnson states that the available power per unit bandwidth of thermal noise is given by Equation 11.25, where K is Boltzmann's constant (1.3805×10^{-23} joules/degree K), and T is the absolute temperature of the source in degrees K.

Noise power $P_n = kT$ watts/Hz (11.25)

Using measurements it has been shown that across the entire frequency range the available power P_a is directly proportional to the product of the system bandwidth B_w and the absolute temperature of the source T, as in Equation 11.26, which may be expressed in dBm at room temperature as in Equation 11.27.

$P_a = kTB_w$ watts (11.26)

$P_a = -174 + 10 \log B_w$ dBm (11.27)

11.8.2 Noise voltage equivalence circuit

A resistor is a good example of a thermal noise source, and therefore a suitable equivalent circuit is an r.m.s. noise voltage generator connected in series with a hypothetically noiseless resistor having the same resistance. If these are connected in series with a load resistor (Figure 11.14) then the maximum noise power will be delivered with $R_{Load} = R$. Therefore the maximum power is given by Equations 11.28 and 11.29, leading to Equation 11.30.

$$P_a = \frac{V_n^2}{4R} \qquad (11.28)$$

$P_a = kTB_w$ (11.29)

$V_n = (4kTB_wR)^{1/2}$ (11.30)

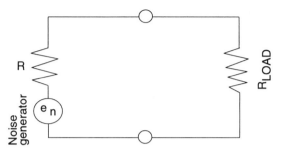

Figure 11.14 Equivalent circuit for resistor noise

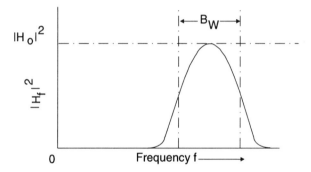

Figure 11.15 Effective noise bandwidth

Reactive elements do not contribute to thermal noise but they do affect its frequency shape.

11.8.3 Effective noise bandwidth

The effective noise bandwidth is defined as the width of a rectangular frequency response curve having a height equal to the the maximum height of the frequency response curve and corresponding to the same total noise power (see Figure 11.15). It is given by Equation 11.31.

$$B_w = \frac{1}{|H_0|^2} \int_0^\infty (H_f)^2 \, df \quad \text{Hz} \tag{11.31}$$

11.8.4 Shot noise

Shot noise is the name given to noise generated in active devices such as valves, transistors and integrated circuits. It is caused by the random varying velocity of electron movement under the influence of externally applied potentials or voltages at the terminals or electrodes.

It is similar to thermal noise in that it has a gaussian distribution and a flat power spectrum. It differs, however in that it is not directly affected by temperature. Its magnitude is proportional to the square root of the direct current through the device and thus may be a function of signal amplitude.

The r.m.s. value of the shot noise current is given by Equation 11.32, where I_a is the average anode current; B_w is the effective bandwidth; and e is electronic charge.

$$I_s = (2 e I_a B_w)^{1/2} \tag{11.32}$$

11.8.5 Partition noise

Partition noise occurs in multi-electrode devices such as transistors and valves. It is due to the current through the device being divided between the various electrodes.

11.8.6 Intermodulation noise

Intermodulation noise is due to the presence of the products of intermodulation. If a number of signals are passed through a non-linear device the result will be intermodulation products that are spurious frequency components. These components may be inside or outside the frequency band of interest for the device.

The non-linearities cause each signal to combine with the other signals in the set, to produce a series of second order sum and difference products, third order products etc.

For most analogue systems with many multiplexed channels, the addition of the very large combination of signals results in an output noise spectrum which is approximately flat with a frequency across a narrow band of about 4kHz.

Intermodulation noise differs from thermal noise since it is a function of the signal power at the point of non-linearity.

11.8.7 Crosstalk

Crosstalk is the unwanted coupling between signal paths. There are essentially three causes of crosstalk:

1. Electrical coupling between transmission media such as between wire pairs on a VF cable system, or a capacitance imbalance between wire pairs in a cable.
2. Poor control of frequency response in an analogue multiplexer system caused by defective filters or poor filter design.
3. Non-linear performance in analogue multiplexer systems.

There are two types of crosstalk:

1. Intelligible crosstalk, where at least four words are intelligible to the listener from extraneous conversations in a seven second period.
2. Unintelligible crosstalk, which is crosstalk resulting from any other form of disturbing effects from one channel to another.

Intelligible crosstalk causes the greatest problem to the listener. This is caused by the fear of loss of privacy or the fact that the listener has great difficulty in differentiating between the party to which he is connected and the interfering circuits.

11.8.8 Impulse noise

Impulse noise consists of short spikes of power having an approximately flat frequency response over the spectrum range of interest. It is considered to be a voltage increase of 12 dB or more above the r.m.s. noise for a period of 12 ms or less.

This type of noise may be caused by a number of sources such as the switching of relays in electro-mechanical telephone exchanges. Although these 'clicks and pops' are annoying to the human ear, it is reasonably tolerant. However impulse noise may cause many serious error rate problems on data or digital circuits.

The effects of impulse noise can be alleviated by the use of a wide band clipping circuit followed by a band limiting filter. This procedure first reduces the amplitudes of the spectral components and then reduces the number of components.

11.8.9 Flicker noise

Flicker noise is sometimes known as low frequency ($1/f$) noise or excess noise. It is because of its unusual increase at very low frequencies that it is also called low frequency noise.

Its cause is associated with contact and surface irregularities in cathodes of valves and semiconductors and it appears to be caused by fluctuations in the conductivity of the medium. Because of the advances in cleaning and passivation techniques, which are employed during component manufacturing processes, a good device will exhibit negligible flicker noise above 1kHz.

11.9 Noise units and measurements

11.9.1 Terms of measure

The measurement of noise is an effort to characterise a complex signal. The noise measurement in telephone channels is further complicated by the subjective nature of the annoyance caused to users, rather than the absolute magnitude of the noise power. A noise measuring set must therefore measure the subjective interference effect as a function of the frequency as well as magnitude and give the same readings for different types of noise which cause equal interference.

The value of 10^{-12} watts or -90dBm is an arbitrary noise reference value. The markings on a noise meter are in decibels and measurements are expressed in dB above reference noise dBrn.

A 1000Hz tone having a magnitude of -90dBm will give a reading of 0dBrn regardless of weighting, but all other measurements must specify the type of weighting.

11.9.2 Equivalent noise resistance

In a circuit using valves or transistor devices shot noise is produced. For noise calculation purposes this shot noise can be thought of as the thermal noise produced by an 'equivalent resistor'.

For a triode the hypothetical resistor is inserted in series with the grid and its value is given by Equation 11.33, where g_m is the mutual conductance of the triode.

$$R_{eq} = \frac{2.5}{g_m} \tag{11.33}$$

11.9.3 Noise temperature

The thermal concept can be applied to other types of noise sources such as diodes and noise tubes. The noise temperature of such a device is the temperature of a thermal noise source which produces the same available noise power as the device under consideration regardless of the actual physical temperature of the device.

The concept of noise temperature is not confined to noise sources. For example the noise power measured at the output of an amplifier or antenna can be expressed in terms of the equivalent noise temperature. In the case of an antenna the noise is due to radiation from objects on earth as well as in outer space and the antenna noise temperature is used.

Another concept which is used is excess noise temperature T_x which is the difference between the noise temperature of the source T and the noise temperature $T0$ of a thermal noise source at room temperature (17° C or 290° K). (Equation 11.34.)

$$T_x = T - T_0 \tag{11.34}$$

Many commercially available noise sources with resistive terminations are calibrated in these terms. In use care must be taken to correct for actual temperature in such devices.

11.9.4 Effects of noise on speech telephony

Noise is a complex signal and the measurement of its effects in telephone channels carrying speech is made more difficult by the fact that, to the telephone user, it is the annoyance that is caused by noise that is important rather than the absolute magnitude of the noise power. The subjective interfering effect of noise is a function of frequency as well as magnitude and may be the combination of a number of dissimilar forms of noise.

The results of many experiments involving telephone users have shown that, within the normal telephone frequency range of 300Hz to 3400Hz, lower frequencies cause less annoyance than higher frequencies. It has also been shown that if simultaneous noises are present then the effect to the user is additive on a power basis, i.e. if tones of an equal interfering effect are used a meter indication would be 3dB higher.

A further effect to be noted is that it has been shown by experiment that the human ear does not fully respond to sounds shorter than 200ms and therefore does not fully appreciate the true power.

11.9.5 Line weighting for channel noise

In order to conduct measurements of channel noise it is necessary to recognise the effects of annoyance to the telephone user of frequency. The results of experiments have produced a number of line weighting curves with respect to frequency. This weighting must take account of the relative annoyance of tones, both in the presence of speech and in its absence, with respect to a tone of a specified frequency.

This weighting also must take account of the effects of attenuation caused by the telephone handset. Thus standard line weighting networks have been produced which take account of these factors and provide a frequency shaping effect which matches that of a telephone user and telephone handset.

11.9.6 Psophometric

Psophometric line weighting is the standard produced by the ITU-T. It is defined as the noise measured on a telephone line using a psophometer, which is an instrument for measuring channel noise and includes a weighting network. For most purposes it is sufficient to assume that the psophometric weighting of 3kHz white noise decreases the average power by about 2.5dB.

The reference frequency specified at the point of minimum attenuation in the voice channel is 800 Hz. The full specification is provided in the ITU-T G series Recommendations.

It is common to refer to average noise power delivered to 600 ohms instead of r.m.s. noise voltage, this power is expressed in picowatts.

The noise units used are picowatts psophometric (pWp) or decibels psophometrically weighted (dBmp), as in Equations 11.35 and 11.36, for noise which is flat from 0kHz to 3kHz.

$$pWp = \frac{(\text{psophometric mV}) \times 10}{600} \quad \text{pwatts} \qquad 11.35)$$

$$dBmp = 10 \log (pWp) \qquad (11.36)$$

11.9.7 FIA C-message noise units

In North America there are two forms of line weighting that are encountered, FIA and C-message.

FIA line weighting is so called because it is based upon the FIA handset developed by Western Electric Company. It uses a reference frequency of 1000Hz and a reference power level of –85 dBm. The unit of noise measurement for FIA weighting is dBa (dB adjusted). FIA line weighting is being phased out.

The second line weighting, which is now the preferred standard in North America, is C-message line weighting which is based upon a more sensitive handset. It uses a reference frequency of 1000Hz and a reference power level of –90dBm. The unit of noise measurement used for C-message weighting is dBrnC (dB above reference noise C-message).

11.9.8 Noise measurement instruments

Instruments for measuring noise are usually built specially for the purpose because of the random nature and considerable fluctuations of the noise voltage.

11.9.8.1 *Voltmeters*

The use of an a.c. voltmeter to measure noise voltage is possible under limited conditions. However, a large amount of fluctuation would be seen due to the random nature of the noise voltage. If a rectifying a.c. voltmeter is used then form factor correction should be used for the measurements. This requires knowledge of the character of the noise to be measured.

A voltmeter will indicate the total r.m.s. voltage of the noise if the voltmeter has a frequency response greater than the bandwidth of the noise spectrum to be measured. If the voltmeter has a frequency response less than that of the bandwidth of the noise then the reading will be proportional to the noise within the bandwidth of the meter.

When the noise to be measured is in a narrow frequency band within a broadband spectrum, then a filter of the required bandwidth must be used between the voltmeter and the point to be measured. It is also common to use frequency shaping, in addition to band limiting, which is known as noise weighting.

11.9.8.2 *Noise meter*

The measurement of noise in a system or device is often performed using a noise meter. This is actually an a.c. voltmeter preceded by a filter which is tunable or switchable over a range of frequencies much greater than their noise bandwidths. In most cases these meters will have a fixed input impedance and are often calibrated in terms of power rather than voltage.

In the case of an average reading voltmeter form factor correction must be applied.

11.9.8.3 *Impulse noise counter*

Impulse noise is voltage spikes on the line which, although tolerable on voice circuits, will cause errors on circuits used for data transmission.

To measure these noise spikes requires a weighting network, a rectifier, a threshold detector and a counter to record the events above the threshold. A means of measuring time is also provided so that a count may be recorded for a fixed time interval.

A typical impulse counter for the voice frequency band has a threshold adjustable in 1dB steps from 40dBrn to 99 dBrn, with a choice of terminating (600 ohms) or bridging (high) input impedance. The timer can be set in 1 minute intervals up to 15 minutes, a 5 minute interval being standard for message circuits. On some instruments there are several counters in order to make counts at different threshold levels simultaneously, which allows the measurement of the distribution of impulse magnitudes.

11.10 Signal to noise ratio

11.10.1 Definitions

Signal to noise ratio is the ratio of the required signal power to the noise power, as in Equation 11.37.

$$S/N = \frac{\text{Required signal power}}{\text{Noise power}}$$
$$= 10 \log \frac{\text{Required signal power}}{\text{Noise power}} \quad \text{dB} \qquad (11.37)$$

In some systems it is useful to compare the input signal to noise ratio $\frac{S_i}{N_i}$ with the output signal to noise ratio $\frac{S_o}{N_o}$.

Signals and noise 293

The performance of a system may be judged in terms of the signal to noise ratio, and for the best results it must be as large as possible.

The theoretical maximum data rate of a transmission medium is related to the signal to noise ratio and can be determined using a formula attributed to Shannon and Hartley. This formula, known as the Shannon-Hartley Law is given in Equation 11.38, where C is the data rate, B is bandwidth in Hz, and N is the random noise power in watts.

$$C = B \log_2 (1 + \frac{S}{N}) \quad \textit{bits per second} \tag{11.38}$$

11.10.2 Effect of amplification

The effect of amplification upon the signal to noise ratio is to degrade it at the output of the amplifier as compared to its input. This is caused because the noise is amplified as well as the the signal and in addition the amplifier itself will introduce noise into the system, usually thermal noise and/or shot noise.

A term often used in connection with amplifiers is 'equivalent noise input' which is defined as the amount of noise that would be required to be present at the input of a completely noiseless amplifier in order to produce the same noise level at the output as that of a practical amplifier.

From this definition it follows that the output noise of an amplifier is equal to the input noise plus the gain of the amplifier in decibels. (See Figure 11.16 and Equation 11.39.)

$$N_{out} = N_{in} + G \quad \text{dB} \tag{11.39}$$

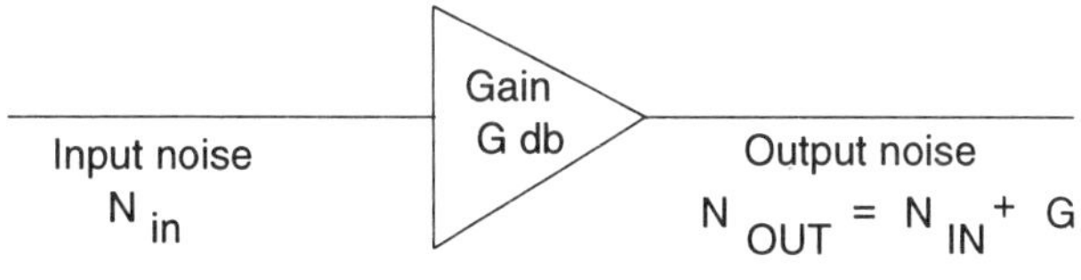

Figure 11.16 Output noise in terms of input noise

294 Signal to noise ratio

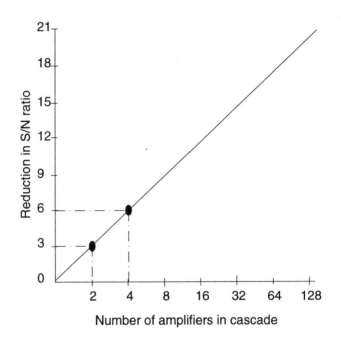

Figure 11.17 Effect on S/N of multiple amplifier connections

11.10.3 Effect of tandem connections

The doubling of noise power, by adding two identical amplifiers in tandem with equal input levels, increases the noise power by 3dB. If the input level remains the same and the noise power level is increased by 3dB, then the signal to noise ratio is reduced by 3dB.

Thus every doubling of the number of amplifiers will show a 3dB reduction in the signal to noise ratio. Figure 11.17 shows the effect of adding more amplifiers upon this ratio. However it should be noted that this will only hold true for amplifiers of identical noise figures. A similar graph may be produced for other combinations of amplifiers.

11.11 Noise factor

11.11.1 Definition of noise factor (noise figure)

The noise factor, usually of an amplifier, is defined as the ratio of the signal to noise ratio at the input to the signal to noise ratio at the output of the amplifier stage. It indicates the 'noisiness' of the amplifier, as in Equation 11.40.

$$\text{Noise factor} = \frac{\text{Signal noise ratio at input}}{\text{Signal noise ratio at output}} \tag{11.40}$$

As the units on the top and bottom of this equation are the same then it will be seen that the noise factor is just a number.

It is convenient to work in decibels and to this end the term noise figure is used, which is the noise factor expressed in decibels, as in Equation 11.41.

$$\text{Noise figure} = 10 \log (\text{Signal noise ratio at input}) \\ - 10 \log (\text{Signal noise ratio at output}) \tag{11.41}$$

It is seen from this equation that the noise figure is the number of decibels by which the signal to noise ratio is degraded by an amplifier.

11.11.2 Noise factor in terms of equivalent noise resistance

As has already been stated, for calculation purposes the shot noise produced by valves and transistor devices may be represented by the the thermal noise produced by an 'equivalent resistor'.

The equivalent resistance and the noise factor relationhip is: the higher the equivalent resistance the higher the noise factor or noisiness of a device. For triode valve circuits the relationship is given by

Equation 11.42, where F is the noise factor, R_{eq} is the equivalent resistance, and R_s is the resistance in series with the input supply.

$$F = 1 + \frac{R_{eq}}{R_s} \tag{11.42}$$

11.11.3 Noise factor and noise temperature

These measures are used to indicate the amount of noise generated in a network. In both cases the higher the value the more noise is generated in the network.

The relationship between noise factor and noise temperature are given by Equations 11.43 and 11.44, where F is the noise factor and T is the noise temperature in degrees Kelvin.

$$T = (F-1)\,290 \tag{11.43}$$

$$F = \frac{T}{290} + 1 \tag{11.44}$$

For low noise systems, where F is small, it is more convenient to use T for calculations which is larger in both range and magnitude.

11.11.4 Input noise in cascaded amplifiers

The concept of equivalent input noise of an amplifier may be used to calculate the noise level in cascaded amplifiers. It may be shown that the noise level at the input of the second amplifier is given by Equation 11.45.

$$N_{total} = N_{in} + 10 \log 2 \tag{11.45}$$

In a similar manner, for system with x amplifiers, the noise input at the xth amplifier (N_x) will be given by Equation 11.46.

$$N_x = N_{in} + 10 \log x \tag{11.46}$$

11.11.5 Noise factor measurement

The most convenient method of measurement of noise factor uses a noise diode, as in Figure 11.18, as this method does not require the bandwidth of the system to be known. The system bandwidth is very difficult to measure to any degree of accuracy.

A noise diode produces mainly shot noise, which may be adjusted by varying the input current to it. It produces a broad spectrum of frequencies up to at least 5MHz. The input resistance of the active network is R_g, the output power is P_o, and the noise factor is F. The following method is used. First measure P_o when the diode current and the attenuation are zero. The diode current is then adjusted to give the same power reading with an attenuation of 3dB (a doubling of the output power). By using the same meter reading any discrepancies caused by the meter are avoided. Equation 11.47 can be obtained, and this reduces to Equation 11.49 if Equation 11.48 holds.

$$F = \frac{e\, I_a R_g}{2\, k\, T} \tag{11.47}$$

$$\frac{e}{2\, k\, T} \approx 20 \tag{11.48}$$

$$F = 20\, I_a\, R_g \tag{11.49}$$

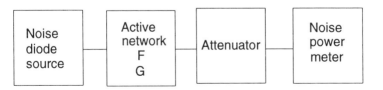

Figure 11.18 Method of measurement of noise factor

11.12 Noise waveforms

11.12.1 Mathematical model

Thermal noise is a combination of a very large number of random events and satisfies the conditions of gaussian distribution, as in Figures 11.19 and 11.20, and Equation 11.50.

$$\rho(V) = \frac{1}{\sigma_n \sqrt{2\pi}} \exp\left(\frac{-V^2}{2\sigma_n^2}\right)$$
$$= \frac{1}{\sigma \sqrt{2\pi}} \int_{-\infty}^{V} \exp\left(\frac{1}{2}\frac{x^2}{\sigma_n^2}\right) dx \qquad (11.50)$$

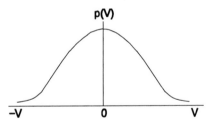

Figure 11.19 Gaussian density function

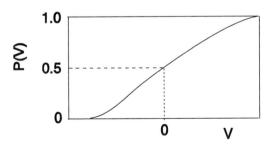

Figure 11.20 Gaussian distribution function

11.12.2 Band limited white Gaussian noise

In most practical applications where white noise is produced it will be passed through a band limiting filter. Filtering uncorrelated white noise produces correlated band limited white noise or coloured noise. The autocorrelation function is given by Equation 11.51.

$$R_\tau = N_o B \frac{\sin(2\pi B_\tau)}{2\pi B_\tau} \tag{11.51}$$

11.12.3 Narrowband Gaussian noise

Narrowband gaussian noise is called Rayleigh noise and it is defined as narrowband if the noise bandwidth is small compared to the midband frequency. It has the appearance of a sinusoidal carrier at the midband frequency, modulated in amplitude by low frequency wave whose highest frequency is dependent on the bandwidth of the noise. (See Figures 11.21 and 11.22.)

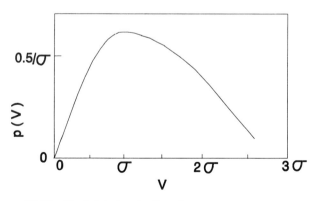

Figure 11.21 Rayleigh density function

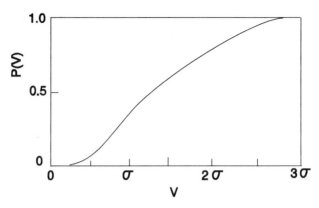

Figure 11.22 Rayleigh distribution function

11.13 Bibliography

Bell (1979) *Transmission Systems for Communications*, Bell Telephone Laboratories.

Freeman, R.L. (1989) *Telecommunication System Engineering*, John Wiley.

Lee, E.A. et. al. (1990) *Digital Communications*, Kluwer Academic Publications.

Slater, J.N. (1988) *Cable Television Technology*, John Wiley.

Index

A-law (compander), 195
Adaptive quantiser, 195
ALOHA models, 160-61
Amplification, 293
   cascaded amplifiers, 296
   tandem connections, 294
Analog-to-digital
   converter (ADC), 193
Anti-jamming, 257
Aperiodic signals, 273
Argand diagram, 9
Arithmetic mean, 51-2
Arithmetic series, 31
Arrival and departure
   distributions, 136-8
Attenuation distortion, 279, 282
Audibility *see* Threshold of
   audibility
Authentication, 253
Average, 51-3
   dispersion from, 53-5

Band limiting, 193, 291
Bandwidth, 268
   effective noise *see*
   Effective noise bandwidth
Baud, 262
Bayes' theorem (probability),
   61-2, 169, 175

BCH (Bose Chaudhure Hocquengherm) codes, 245-6
   see also Reed-Solomon codes
Berklekamp/Massey
   algorithm, 246, 247
Binary channel, 231
   symmetric, 181-3, 231
Binomial series, 31
   probability distribution, 62-3
Birth-death process,
   queuing models, 133-6, 199
Bit error rate (BER), 231
Bits, 262
Block coding, 228
   2-D parity codes, 236-7
   concatenated codes, 239
   extended codes, 238
   Hamming codes, 236
   Hamming distance, 233-4
   hard decision decoding,
     234, 235
   interleaved codes, 239
   linear, 232-3
   other codes, 237
   performance, 233-4
   shortened codes, 237-8
   single parity checks, 231-2
   soft decision, 234-5
   see also Cyclic codes

Boltzmann, information
    theory, 169-70, 172
Brouwer fixed point theorem, 209
Burst errors, 230, 247-8

C-message line weighting,
    290-91
Calculus
    derivative, 21, 22-6
    double integral, 28
    integral, 12, 22, 22-6
    integral transformations, 14
    maxima and minima, 21-2
    numerical integration, 27-9
    reduction formulae, 27
    standard substitutions, 27
    vector, 29-30
Cauchy-Riemann equations,
    10-11
Cauchy's integral
    theorem, 11, 113
CDMA applications, 257
Channel capacity, 179-83
    see also Shannon-Hartley
        theorem
Channel coding, 227
Channel matrix, 178
Channel noise, line
    weighting, 289
Channels
    binary symmetric, 181-3
    compound, 192
    continuous, 192
    Gilbert model, 189-90
    noiseless, 186-7
    noisy, 185-7
    trapdoor, 188-9

Charts, 49
Check digits *see* Parity digits
Chi-square test, 73-6
Ciphertext, 253
Coder-decoder (CODEC), 195
Coders, cyclic *see* Cyclic coders
Codes *see* Error control coding
Codewords, 232-3
Coding *see* Error control coding
Coding gain, 231
Coefficient of variation, 55
Cofactor, determinant, 47
Combinations, 56-7
Compander, 194-5
Complex signals, 275
Complex variables, 9-10
Compressor, 194
Concatenated coding, 239
Constant ratio codes, 237
Continuous systems theory, 107
Convolutional codes, 228, 248-50
    performance, 252-3
    Viterbi decoding, 250-52
Coordinate systems
    Cartesian, 13
    cylindrical, 13
    matrix transformation, 45
    spherical polar, 13
Correlation, 59-60
    significance, 76
Crosstalk, 286
Cryptosystems *see*
    Encryption, specific systems
CSMA (Carrier Sense
    Multiple Access), 162-4
Current ratio, 20
Cyclic coders, 242-3

Cyclic codes, 239
  BCH codes, 245-6
  implementation 242-5
  mathematics, 240-42
  other, 247-8
  Reed-Solomon codes, 246
  shortening, 248
Cyclic decoders, 243-5

Data
  presentation, 49-51
  truncation, 113-18
Data Encryption Standard (DES), 255
Data windows, 116-18
  flow control, 154
De Moivre's theorem, 8
Decibels, 19, 20
Decimation in time (DIT) graph, 100, 101, 102, 103
Decoders, cyclic *see* Cyclic decoders
Decoding
  algorithms, 246
  hard decision, 234, 235-6
  soft decision, 234-5
Degrees of freedom, probability, 73-6
Delay distortion, 280
Delay performance, 121
Delta modulation, 195
Determinants, 45-6
  properties, 46-7
Deterministic signals, 274
Dibit, 262
Differential mean value theorem, 28
Differential pulse code modulation (DPCM), 195-6
Digital transmission, 266-8
  distortion, 282
  see also Time invariant digital systems
Dirac delta functions, 39, 93-4, 115
Direct sequence spread spectrum (DSSS), 256-7
Dirichlet condition, 84
Dirichlet kernel, 115
Discrete channels
  with memory, 188-92
  memoryless, 177-9
Discrete Fourier transforms (DFT), 94-7
Distortion, 279-82
  digital signals, 282
Dual error detection (DED), 236
Dynamic Alternative Routeing (DAR), 218, 220-22
  BT network, 222-4
Dynamic Non-Hierarchical Routeing (DNHR), 217, 21819
Dynamic routeing strategies *see* Routeing strategies, dynamic
Dynamically Controlled Routeing (DCR), 21718

Effective noise bandwidth, 284
Eigenvalues and Eigenvectors, 44-5
Embedded Markov chain, 139-40

Encoding, source, 183-5
Encryption
  applications, 253
  principles, 253-5
  specific systems, 255
Energy signal, 272
Entropy, 166-7
  of finite schemes, 173-4
Equation solutions
  bisection method, 16
  fixed point iteration, 17-18
  linear, 47-8
  linear simultaneous, 42
  Newton's method, 18
  quadratic 15-16
  regula falsi method, 16-17
Equilibrium models
  distribution network, 153
  queue distribution, 135, 138, 149
  queuing, 121
Equivalent noise input, 293
Erlang, A. K., 197
Erlang bound, network, 216-17
Erlang fixed point, 205-10, 212, 213, 221, 224
Erlangian distribution, 126
Erlang's loss formula, 200, 213
Error control coding (ECC)
  arithmetic, 229-30
  choice, 231
  feedback, 229
  feedforward, 229
  need, 227
  principles, 227-8
  types, 228
  see also Block coding; Convolutional coding
Error correction, codeword, 234
Error pattern detector, 245
Errors, types, 230-31
Euler's relation, 8
Excess noise *see* Flicker noise
Expander, 194
Exponential distribution, 67-8
Exponential form formulae, 8

Fast Fourier transform (FFT), 98-100
Feedback error correction, 229
Feedforward error correction (FEC), 229
FIA line weighting, 290
Fisher, R. A., 174
Flicker noise, 287
Fourier analysis
  background, 78-84
  generalised expansion, 84-90
  series expressions, 34-5
  signal representation, 275-8
  see also Fourier transforms
Fourier transforms, 37-9, 90-93
  discrete, 94-7
  fast, 98-100
  inverse discrete, 98
Frequency, 263
  fundamental, 265-6
  representation of distributions, 49-51
  shaping, 291
Frequency domain, signal, 274

Frequency hopping spread spectrum (FHSS) system, 257-8
Frequency response, 112

G/M/m queue, 143-6, 147
Galois field arithmetic (GF(M)), 229-30, 240
Gamma distribution *see* Hyperexponential distribution
Gaussian density function, 298
Gaussian distribution function, 298
Gaussian noise, 185, 282
  band limited, 299
  narrowband, 299
Geometric mean, 52
Geometric series, 31
Gibbs' oscillations, 84
Gilbert model, 189-90
Global balance equations, 133
Golay code, 237
Gold sequence, 257
Graphs, 49

Hadamard codes, 237
Hamming codes, 236
Hamming distance, 233-4
Harmonic distortion, 280-81
Harmonic mean, 52-3
Harmonics, waveform, 266
Hartley, R. V., 174
Histograms, 49, 50
Hyperbolic functions, 9
Hyperexponential distribution, 126-7

Hypothesis testing, 72-3

Ideal observer scheme, 187
Implied costs, network models, 212-14, 214-15
Impulse noise, 286-7
  counter, 292
Information capacity, of store, 171-2
Information theory, 166-7
  see also Mutual information; Self information
Integer series, 33-4
Integral mean value theorem, 29
Integrals *see under* Calculus, integrals
Interference, effect of, 230
Interleaving/interlacing, 239
Intermodulation
  distortion, 281
  noise 285-6
International Alphabet 2 (IA2) code *see* Murray Code
Inverse discrete Fourier transforms (IDFT), 98
ITU-T Recommendation G series, 290
ITU-T Recommendation V.22 (dibit operations), 262
ITU-T Recommendation V.29, 262

Jacobian function, 28
Johnson noise *see* Thermal noise

Kasami sequence, 257
Kendall Queuing notation, 125-8

Key management, cryptosystem, 254-5
Kolmogorov, information theory, 170

Laplace transforms, 39-41, 107, 138-9
Laplace's equation, 14-15
Laplace's law of insufficient reason, 169
Laurent's series, 32
Learning automation schemes, 219-20
Least busy alternative (LBA) strategy, 224-5
Legendre polynomials, 84, 86
L'Hopital's rule, 141-2
Line weighting
  C-message, 290-91
  for channel noise, 289
  FIA, 290
  psophometric *see* Psophometric line weighting
Linear block codes, 232-3
Linear equations, numeric solution, 47-8
Linear Feedback Shift Registers (LFSR), 242
Linear predictive coding (LPC), 196
Linear recursive operators, 110
Linear simultaneous equations, 42
Little's law (queuing), 124-5, 134-6
Local balance equations, 133, 134

Low frequency noise *see* Flicker noise

M-ary digit, 229
µ-law (compander), 194-5
M/G/1 queue, 139-43, 146-7
Maclaurin's series, 32
Markov processes, 131-3, 134
  chains 131, 139, 161
Mathematical signs and symbols, 1-4
Matrix arithmetic, 42-3
  adjoint, 43
  eigenvalues and eigenvectors, 44-5
  inverse, 43-4
  orthogonality, 44
  product, 43
  transpose, 43
Max-flow bound, network, 215-16
Maximum likelihood decision scheme, 188
Maxwell, 172-3
Mean deviation, 54
Mean squared error (MSE), 80-81
Median, 52
Meggitt decoder, 244, 245
Memoryless property, exponential, 130
Method of least squares, 18-19
Minor, determinant, 47
Mode, 52
Modulation, 193
  see also Delta modulation
Modulo-2 arithmetic, 230
Morse Code, 270

Multi-access channels, 159-64
Murray Code, 270
Music signals, 268-9
Mutual information, 174-7

Network models, 124
  application of implied costs, 214-15
  bounds, 215-17
  dimensioning, 214
  Erlang fixed point, 205-10
  implied costs, 212-14
  see also Single link network model
Networks, queuing *see* Queuing networks
Newton's method, 18
Noise, 282-7
  cascaded amplifiers, 296
  effects on speech telephony, 289
  equivalent resistance, 288, 295-6
  line weighting, 289
  measurement instruments 291-2
  measurement terms 287-8
  random 282-3
  waveforms, 298-300
  weighting *see* Band limiting
Noise diode, 297
Noise factor, 295
  equivalent noise resistance, 295-6
  measurement, 297
  and noise temperature, 296
Noise figure, 295

Noise meter, 291
Noise reference value, 288
Noise temperature, 288-9
  and noise factor, 296
Noise voltage equivalence circuit, 283
Non-periodic signals *see* Aperiodic signals
Normal distribution, 64-6
Nyquist rate, 193

Ogives, 49, 51
Orthonormal sets, 82-3

Packet switched networks, 124
Parity check matrix, 233
Parity codes, 2-D, 236-7
Parity digits, 228
Parseval's lemma, 39, 81, 88, 93
Partition noise, 285
Pearson coefficient (skewness), 55-6
Periodic signals, 273
Periodic time, 263
Permutations, 57
Peterson's direct solution, 246, 247
Phase, alternating current, 264
Phase difference, 264
Phase distortion, 280
Plaintext, 253
Poisson process, 63, 130, 139, 161
Pollazcek-Kinchine formula, 146
Power ratio, 20
Power series for real variables, 33
Power signal, 272

Probability, 60-62
  distributions, 62-70
  information theory, 167-71
Probability flux, 133
Pseudo random binary sequence (PRBS) generator, 254
Pseudo-noise (PN) signal, 256
Psophometric line weighting, 290
Pulse Code Modulation (PCM), 193, 195-6
Pulse wave, 37, 277

Quantisation, 193-4
Quartiles, 54
Queuing models
  birth-death process, 133-6
  fixed point networks, 156-9
  general customer routes, 147-56
  Markov processes, 131-3
  networks, 124
  Poisson process, 130
  reversed networks, 151
  state of network, 149
Queuing problems, 121-4
  exponential distribution, 128-9
  see also Kendall Queuing notation; Little's law

Radar, 271
Radio systems, 270
Random errors, 230
Random noise, 282-3
Random routeing, 207
Random signals, 274
Range (series), 53-4
Rank correlation coefficient, 60

Rayleigh density function, 299
Rayleigh distribution function, 299-300
Rayleigh noise *see* Gaussian noise, narrowband
Real Time Network Routeing (RTNR), 225
Rectified sine wave, 35-6
Reed-Solomon (RS) codes, 246-7
Regression, 57-9
Routeing strategies
  decentralised adaptive, 214-15
  dynamic, 217-218
  dynamic alternative, 218, 220-24
  dynamic non-hierarchical, 217, 218-19
  least busy alternative, 224-5
  real time network, 225
RSA encryption algorithm, 255

Sampling, data, 70-72
Sawtooth wave, 37, 278
Scatter diagram, 57, 58
Self information, 174-7
Sequences, discrete, 93-4
Shannon, C., 174
Shannon entropy, 170, 174
Shannon-Hartley law, 185-6, 293
Shannon's first theorem (source encoding), 185
Shot noise, 285
Signal frequency, 260
Signal to noise ratio, 292-3
  effect of amplification, 293
  effect of tandem connections, 294

Signal types
  radar 271
  radio 270
  telegraphy, 270
  telephone, 269
  television, 270-71
  voice and music, 268-9
Signalling, 260
Signals, 260
  alternating current, 261
  classification, 271-4
  direct current, 260-61
  distortion, 279-82
  representation, 274-8
Simpson's rule, 28
Sine/cosine signals, 275
Single error correction (SEC), 236
Single error detecting (SED) code, 236
Single link network model
  Erlang's loss formula, 198-202
  trunk reservation, 202-205
Skewness, 55-6
Slotted ALOHA protocol, 123
Small angle approximations formulae, 6
Sojourn time, 156
Source encoding, 183-5, 228
  theorem *see* Shannon's first theorem
Spectral content, signal, 114
Spectral leakage, 96, 116
Speech telephony, effects of noise, 289
Spherical triangle formulae, 7-8
Spread spectrum systems
  applications, 256
  direct sequence, 256-7
  frequency hopping, 257-8
Square wave, 36, 276-7
Standard Array, 236
Standard deviation, 55
Standard error of estimation, 59-60
Standard error of the mean, 70-72
Standard integral forms, 12
'Sticky random' strategy *see* Dynamic Alternative Routing (DAR)
Stirling's approximation, information theory, 170
Store, information capacity, 171-2
Store and forward networks *see* Packet switched networks
Student t test of significance, 76
Synchronisation distortion, 282
Syndrome, codeword, 233
Systematic coding, 228
Szilard, 172, 173

Taylor's series 11, 32
Telegraphy systems, 270
Telephone systems, 269
Teleprinter systems, 270
Teletraffic, 197
  dynamic routing, 217-25
  network models 205-17
  single link models 198-205
Television systems, 270-71
Tests of significance, 72-6
Thermal noise, 282-3
Threshold of audibility, 269

Threshold of feeling, 269
Time domain, signal, 274
Time-invariant digital systems
   convolution summation property, 109-13
   delay property, 108-109
   linear, 104-108
   linearity, 108
Transfer function, 107, 111
Transform methods
   arrival and departure distributions, 136-8
   results, 146-7
   see also G/M/m queue; M/G/1 queue
Transformation of integrals, 14
Transition probability matrix, 178
Trapezoidal rule, 27-8
Trellis diagram, 250
Triangle solutions formulae, 6-7
Triangular wave, 36, 278
Tribit, 262
Trigonometric formulae, 4-6
Trigonometric values, 6
Trunk reservation, 202-205, 210
Twiddle factors, 94

Vector calculus *see* Calculus, vector

Viterbi decoding, 250-52
Voice signals, 268-9
Voltage ratio, 20
Voltmeter, for noise measurement, 291
Von Neumann, J., 174

Waveform, 35-7, 262-3
   amplitude, 264
   encoding techniques, 196
   Fourier series representation, 275-8
   frequency, 263
   noise, 298-300
   pulse train, 277
   saw tooth, 266, 278
   square, 266, 276-7
   triangular, 278
Wavelength, 265
Weibull distribution, 68-70
White noise *see* Thermal noise
Windowing *see* Data windows

z-transform 105-106
   common pairs, 106
   inverse, 113
   properties, 111
Zeros, poles and residues, 11-12